WE'LL ALWAYS HAVE LINSANITY

WE'LL ALWAYS HAVE LINSANITY

STRANGE TAKES FROM THE STRANGEST SEASON IN KNICKS HISTORY

JIM CAVAN, MIKE KURYLO, SETH ROSENTHAL,
AND ROBERT SILVERMAN

with Jake Appleman, Jamie Concepcion, Jared Dubin, Dan Litvin,
and Jamie O'Grady

Introduction by Will Leitch

99

Lowell and San Francisco

99: THE PRESS
P.O. Box 956
Lowell, MA
http://99thepress.com

Portions of this work reprinted from *KnickerBlogger*, Twitter, *The New York Times, Hardwood Paroxysm*

ISBN-13: 978-0-9882662-0-9

TABLE OF CONTENTS

---

1 During an interview in the Las Vegas Summer League in his rookie season, when asked what he brought to the court, Mr. Douglas famously described himself in the third person, saying that he'd "Do What Toney Douglas Do." The Acronym, "DWTDD," was popularized in the Knick blogosphere shortly thereafter.

2 If you don't know where the nickname Humpty is from, get yourself to a Google and type in "The Humpty Dance." Then look at a picture of Balkman. Laugh heartily.

BLOGGEROLL CALL

JAKE APPLEMAN (@JakeAppleman) is a frequent contributor to *The New York Times* and *NBA.com*, as well as a senior writer for *SLAM* magazine. He spent much of last season standing next to Mike D'Antoni or Mike Woodson in media scrums, rushing to be by their side so as to not strain his arm while holding a tape recorder aloft. Jake lives in an apartment rigged by a psychopath to detonate if ever Jodeci isn't playing on the stereo. It's like a *Speed* situation, you see, but with Jodeci.

JIM CAVAN (@JPCavan) is a contributor at *KnickerBlogger.Net.* His work has also appeared at *The New York Times* Off the Dribble blog and *ESPN.com.* When not writing about basketball, he serves as director of media for Green Alliance, some New Hampshire hippie group. Jim lives in the most barren stretch of the White Mountain wilderness, where he presides over a colony of feral cave children who regard him as a deity.

JASON CONCEPCION (@netw3rk) is a lifelong Knicks fans who cried during the Charles Smith game. He runs a popular sports-centric Twitter account and is a contributor at *SB Nation.* Jason lives in New York with his master—a giant corgi who occasionally feeds and bathes him.

JARED DUBIN (@JADubin5) is co-editor in chief of *Hardwood Paroxysm* and founder of *Hoop Chalk*, both of which are part of ESPN's TrueHoop Network. He previously founded the mildly popular NBA blog *Outside the Arc* and wrote for the now-defunct and always-unknown basketball Web site *Swoosh Nation.* Jared lives happily inside an industrial, building-sized refrigerator. He sleeps in the vegetable crisper.

DAN LITVIN (@KnicksFanBlog) is an attorney, Twitter antagonist, and the co-editor of *The Knicks FanBlog*). Since 2008, he has published hundreds of entries, some of which have been acknowledged in outlets such as *Grantland* and *Newsday.com.* Dan lives in and has never left the man-sized mussel shell whence he was born.

MIKE KURYLO (@KnickerBlogger) is the founder and editor of *KnickerBlogger.net*, a member of ESPN's TrueHoop Network. He also contributes to *The New York Times* Off the Dribble blog. Mike lives as an array of molecules dispersed miles above the Earth's surface that can be instantly summoned in their whole, corporeal state to any place in the world but only by the keeper of a special amulet.

JAMIE O'GRADY (@OpinesMine) has covered the New York Knicks for *The Journal News* (LoHud.com) and contributes NBA content to *The New York Times* Off the Dribble blog. He has appeared as a guest on WFAN and ESPN Radio, and he fancies himself a Fruit Roll-Ups afficianado. Jamie lives with his wife and dozens of children atop a massive helium balloon floating over the Atlantic Ocean.

SETH ROSENTHAL (@seth_rosenthal) is the founder and editor of *Posting and Toasting* and a regular contributor at *New York Magazine*'s Sports Section blog and *SBNation*. He has also made things that appeared on *GQ.com, NBA.com*, *NYKnicks.com*, and *Yahoo! Sports* and in several coloring books of intermediate difficulty. Seth lives in upstate New York with a parrot, four whippets, a sheep, one and a half ducks, a binturong, and a sentient, alcoholic desk lamp.

ROBERT SILVERMAN (@BobSaietta) writes for the ESPN/TrueHoop's *KnickerBlogger.net*. His writing about basketball can also be found at *Hardwood Paroxysm* and *The New York Times* Off the Dribble blog. He also indulges another passion almost as quixotic and *a priori* frustrating as Knicks fandom, the theater, where he has acted, produced, and written over 20 original plays that have been seen in New York and across the globe. Robert lives in the tallest turret of a Norman castle, where he guards the special amulet and hordes access to Mike Kurylo's presence.

DRAMATIS PERSONAE

AMAR'E STOUDEMIRE: A veteran star beset by unfit company, mismanaged injuries, and personal tragedy.

TYSON CHANDLER: A monumental newcomer with legendary defensive ability and greatness made clear by tasty flecks of a 2011 NBA title still stuck in his beard.

CARMELO ANTHONY: A superstar in form and frame of mind, if not always function. The subject of much disdain, in large part because of the heaping ransom it took to secure his services.

JEREMY LIN: A young, Harvard-groomed journeyman with a cult following, a hidden trove of talent, and a remarkably pun-able surname.

LANDRY FIELDS: A charming, polished sophomore looking to build off a surprisingly terrific rookie campaign with a refined jump shot. What could possibly go wrong?

IMAN SHUMPERT: A slick, lovably headstrong rookie whose outstanding physique and work ethic are at odds with his own on-court poise and ligamental fortitude.

STEVE NOVAK: A slender, stiff, mealy-mouthed Midwesterner who is near perfect at a singular hand motion—a motion that happens to be worth three points in basketball.

J.R. SMITH: A lockout-borne mercenary, made in the U.S.A., but assembled in China, summoned back stateside to shoot. And shoot. And shoot.

BARON DAVIS: The fuzzy, bloated, minimally useful husk of a star point guard. Very smiley.

MIKE BIBBY: The pallid, doddering, slightly more useful husk of a star point guard. Very slimy.

TONEY DOUGLAS: A wholesome if uninspiring combo guard whose reputation among fans plummets as quickly as his performance level.

JOSH HARRELLSON: A big, scrappy country boy rookie with secret penchants for shooting threes and ripping vile farts.

JEROME JORDAN: A languid, loping Jamaican big man back to ride the pine after spending what would have been his rookie season playing in Eastern Europe's comedy circuit.

JARED JEFFRIES: A lanky, uncoordinated veteran retread with defensive talents lost on those who value aesthetics.

RENALDO BALKMAN: An aimless, dreadlocked drifter making more money than most of his teammates to not play.

BILL WALKER: An intermittently useful swingman with a quick trigger finger and a healthy distaste for physical exertion.

DAN GADZURIC: Whatever.

INTRODUCTION

BY WILL LEITCH

As far as my Knicks fandom goes, I've always felt like a little bit of a fraud. I grew up in Central Illinois, about an hour-and-a-half from St. Louis, so my sports allegiances were deep, vast and inextricable from my larger personality. I saw my first St. Louis Cardinals game in 1982. I visited the Assembly Hall in Champaign in 1983. I watched the Big Red St. Louis football Cardinals lose a split-season playoff game in 1984. I knew who my favorite team was before I was 10 years old, before I knew much of anything about anything. This is the essence of sports fandom—you are a fan of your favorite team before, really, you're almost anything else. The *only* thing I knew about myself at seven was that I loved the Cardinals. Sometimes I wonder if it's the only thing I know about myself.

But the NBA has always been different. St. Louis didn't have a pro basketball team, and my father didn't care much for the NBA. So even though I grew up enjoying the NBA, I was agnostic, allegiance-wise. I always sort of enjoyed this. The NBA was the only league I watched—all the way down to my sister's softball league for six-year-olds—without a clear, passionate rooting interest in mind. This made for a stress-free sports experience—no matter what happens, you win.

Some people go their entire lives like this, following sports but never emotionally *investing* in them. By the time I reached adulthood, I knew I'd have to get off the fence; to care, for anything to *matter*, you must feel pain. But who? In January 2000, I moved to New York City and got my answer—The Knicks. They were a logical fit. I wanted some sort of connection to this new city, I had no team anyway and besides, hey, they were good. They'd made the playoffs for 14 consecutive years—my Cardinals had made the baseball playoffs just once in that time. It'd be nice to finally cheer for a winner. That's the thing I already was learning about New Yorkers—they demanded winners.

As you already know, I didn't exactly get my expected run of playoff games and championships. But that was OK, because I was so happy to finally have an NBA team to call my own—one I could just go see anytime I wanted to, assuming I could afford it. (Which I couldn't, of course.) I obsessed over the Knicks roster and essentially forgave every misstep of the Isiah Thomas era with good cheer, thinking, "Hey, that's OK! They'll turn it around! I'm a Knicks fan now! Go Knicks! Weee!" About 10 years ago, my mother had a spiritual awakening and decided to join the Catholic Church. There is no greater zealot than a convert, and before you knew it, she was the most Catholic person in town. This was great for her, but for those who had grown up Catholic—with all the guilt and childhood angst that entails—got tired of my mom awfully quickly; it was so much easier for her because she skipped all the hard parts. I had no past with the Knicks to overcome, so I took the losses and the dysfunction and the Michael Doleac in stride the way a lifelong fan couldn't.

Deep down, though, I knew I hadn't earned it, this Knicks fandom. I felt like I hadn't experienced the highs and lows, the joys and the pains, the sheer madness of life as a fan of the New York Knickerbockers. I always felt like a poser.

And then came last year. After last year, I no longer feel like a poser. And after this offseason, I *really* don't feel like one.

The 2012 season, well, that was the one that put hair on my Knicks fan chest. (Knicks Fan Chest. That sounds like some sort of promotion for Queens kindergartners.) This was the year that had *everything.* Wild win streaks. Horrific losing streaks. Coaches resigning. Superstars both being booed and rabidly cheered, often in the same game. Guys punching fire extinguishers; fire extinguishers not punching back. (Yet.) And, above it all—and the reason the 2012 Knicks mattered, and will always matter, no matter what sort of offseason insanity might have commenced—Linsanity, the most glorious, spontaneous expression of Knicks fandom in the Leitch Era. (Note: I am the only person who calls it this. I hope.) Spike Lee said he'd never heard the Garden louder than during Linsanity; Alan Hahn, who knows the Knicks better than anybody, said, "The roar that comes up in this building when this kid does something is unlike anything I can recall." That happened last year. That will not be soon forgotten.

To my eyes, one of the main reasons the Garden exploded for Lin was that he had nothing to do with the Knicks. He was a West Coast guy not tied up with the misery of the Knicks, the Jim Dolan drama, and the Isiah Thomas horrors. This was not like Carmelo "coming home." He was entirely fresh and new to all this. He seemed legitimately surprised by Knicks fans' passion for their team and for his game. He came from nowhere to make this happen, so there was no pressure to "save" the Knicks, because, hell, what has Jeremy Lin ever cared about the Knicks? He was something organic. The Jim Dolan era has been marked by its consistent, bloated peddling of horse manure, and even when he brings in players we like (Carmelo, Stoudemire, Jason Kidd, so on), you can never escape that nagging sense that, somehow, cheering is letting the bad guy win. Lin came out of nowhere and had none of the Dolan baggage, so when he exploded onto the scene it was transformative in a way the Knicks, under Dolan, aren't supposed to be. It was beautiful in a way because it was so un-Knick-like. The Knicks had been handed the most thrilling, unexpected, glorious sports story in years. And he did the impossible. He made you forget, for a brief while, that *these were the Knicks.*

And now he is gone, like that, for stupid reasons we'll probably never understand. It is the belief of this Knicks fan that this era will be known as pre-Lin and post-Lin. I really believe this. So looking back at last season is something we'll be doing a ton of the next decade. So this book is going to come in handy. Because I think it's the season that'll end up changing everything.

And this I know for sure—I have my pain now. I've had my Knicks fan heartbreak. I've earned it now. I have my Knicks Fan Chest Hair.

FIRST QUARTER:
DECEMBER 25, 2011 – FEBRUARY 3, 2012

And so it began.

Once again, after months of wading through a protracted and seemingly interminable lockout, long-suffering Knicks fans were semi-giddy at the potential (a truly burdensome label indeed) of the 2011-12 Bockers. The Knicks possessed their first viable pivotman since Marcus Camby, flanked by a couple of perennial all-stars and an intriguing young backcourt. Naturally, as has been the case for the entirety of the Dolan regime, the team wildly underperformed.

Amar'e Stoudemire looked ground-bound and creaky, a rapidly aging shell of his former self (despite supposedly gaining 20 pounds of muscle and an inch of height during the offseason). Carmelo Anthony started out well, but hampered by a string of minor nagging injuries, struggled to find his normally devastating mid-range game. Toney Douglas was an absolute nightmare as the starting point guard, utterly unable to run Mike D'Antoni's fast-paced sets and bricking enough outside shots to build a medium-sized brownstone on the Upper West Side. Landry Fields' second-half rookie struggles metastasized into a full-blown sophomore slump.

More importantly, they looked as miserable playing the game of basketball as their fans felt watching them. It looked like the 66-game, truncated season would devolve into a joyless dirge. If New York wanted to make the playoffs, especially considering their investment in Anthony, Stoudemire, and Tyson Chandler, they needed to right the ship—and quickly.

Amar'e's take on the era: "Phenomenal."

A psychiatrist's diagnosis of Knicks fans: "Delusions of Baron Davis's grandeur"[1]

1 Baron Davis was coming off back surgery and had clearly peaked as a player four to five years ago.

#1 single in Finland: Lord Est & Mikael Gabriel, "Vuosi Vaihtuu"

Toney Douglas's Google search history: "define point guard"

From the papers: "If somebody wakes up with a cold, he's playing a lot. If not, then we'll see." (Mike D'Antoni on Jeremy Lin, *New York Daily News,* 12/27/11)[2]

Key stats: After the Knicks' first 23 games, Toney Douglas "boasted" a 39.2 true shooting percentage (that's really bad) and 25.0 three-point percentage while logging 22.5 minutes per game, and—saddest of all— 0 DNPs (did not play). In short, if the first two belong to you, the third and fourth shouldn't. This mercifully brief period will heretofore forever be known as *Douglasanity.*[3]

---

2 Jeremy Lin signed with the Knicks on December 27 after being released by the Rockets. The Knicks were worried about its point guard depth after Baron Davis suffered a series of setbacks while rehabbing his injured back, pushing his return date further and further into an unknown future. As a Taiwanese American point guard from Harvard, Lin had gained some notoriety during his rookie season with Golden State (which had also released him to make room for another point guard, second round draft pick Charles Jenkins) but to date, Lin had received scant playing time and achieved little success.

3 Copyright 2012 by Mike Kurylo.

CHAPTER 1
THE UNDEAD LOCKOUT

BY ROBERT SILVERMAN
KnickerBlogger.net

Throughout the summer and heading into the early autumn, there was a serious question as to whether or not there'd be an NBA season at all. Fans were left clinging to tweets from reporters who were staked out in various hotels where league officials and the players' union were meeting and were inundated with myriad labor and revenue questions which, for the most part, they were ill equipped to handle and seemed as enjoyable as extensive, invasive dental surgery. On the evening of November 30, when the article below was written, both sides were coming face-to-face with a self-imposed deadline to come to an agreement. If they failed, it seemed like the season would in fact be cancelled in its entirety.

I really like zombie movies.

Not so much the new ones—the "*28 Days Later*[4] School," if you will—where it isn't actually the undead but a semi-logically explained "rage" virus that has turned regular, not-dead people into berserker-type monsters. I prefer the real maxilla, mainly because the virus zombies tend to run really, really fast. That's scary, yo. Too scary. Fast-moving zombies completely shatter my delusion that when (not if, when) the evil horde arrives, I could weave my way through/away from a flock of brain-cravers because they're staggering around like I used to after a night of pickling myself with vodka. If they're run-

4 *28 Days Later* is a film that depicts a group of survivors making their way in a post-apocalyptic, dystopian world that has been overrun by a "rage virus" which has transformed the populace into a zombie-like horde. Unlike previous iterations, these zombies run very quickly and have not, technically speaking, risen from the dead.

ning a sub-4.4/40-yard dash, I'm lunch.

These days, it seems there are plenty of folks out there who share my predilection—given the success of *The Walking Dead*[5] (Yes, I watch it religiously. If the makers of that fine televisual program read this, I'd prefer more zombies and fewer heartfelt conversations, please. Thank you.), all the Romero[6] reboots, and the un-killable (pun intended) *Resident Evil*[7] franchise. If you'll allow your humble correspondent to get all sociologically speculative n' shizznit, I think the reason zombies have surpassed vampires as our collective current bête noire, is that no matter how horrible life might be when the zombies overrun the planet, at least it'd be…well…simple.

It wouldn't matter what your credit score is. It wouldn't matter if you did or didn't possess the finest brand-name clothes/car/home/job. It wouldn't even matter that the woman you've been flirtatiously texting for weeks finally agrees to go out with you and in the midst of what seems to be a great date, dropped the "I think we're just friends" deuce right in your lap. All of the maddening inconsistencies and vast shades of grey that make up modern life would be gone in an instant. All that would matter is whether or not you can whack the noggin off the undead creature standing in front of you with a cricket bat or a cast-iron skillet or even your clenched fists of fury before it eats you.

That's, if nothing else, simple, right? Back to the true nature of man. Kill or be killed. Survival. The human animal at its essence.

That's what makes the zombie apocalypse so appealing. In the name of intellectual honesty, I'm certainly not the first to have/hold this theory. There's a really sweet (if lengthy) back and forth with more or less the same thesis over at *Grantland* (though I did think the above thoughts before I read that article).

The reason for this possibly overly lengthy, seemingly jejune preamble is because at this hour, the hard-line owners (and it's oh-so-deliciously ironic that we Knicks fans may find yet another reason to loathe Michael $%@#ing Jordan) are thiiiiiis close to blowing the whole durned thing to smithereens. And as I check Alan Hahn's Twitter feed for minute-by-minute updates, I've been trying to figure out why (aside from the obvious "There might be no games" aspect) this is all so enervating, so profoundly upsetting.

5 *The Walking Dead* is a television series on the AMC Network based on the popular graphic novel of the same name that depicts a group of survivors making their way in a post-apocalyptic, dystopian world that has been overrun by a zombie horde.

6 George Romero directed the seminal horror classic, *Night of the Walking Dead,* and many of the sequels. The films all depict a group of survivors making their way through a post-apocalyptic, dystopian world that has been overrun by a zombie horde.

7 *Resident Evil* is a movie franchise based on a popular video game of the same name that depicts a genetically enhanced superhero making her way through a post-apocalyptic, dystopian world that has been overrun by a viral infection that has engendered a race of zombie-like and other monstrous creatures.

At least for yours truly, it's because sports, like the aforementioned zombie paradigm, project the illusion there exists a world beyond the realm of money. Now of course, we all know this is patently false. Half the stories one might read about one's team in any calendar year will inevitably be about free agent X or salary cap room, or—if you like the Mets—how much money your intrepid owners will have to pay Madoff's victims. The game, though, is meant to stand alone as pure combat. The Warrior-Poet. Effing Spartacus in the ring. Me v. You. Who's better? I dunno, let's play and see. Not, "Let's go through your 990 return for fiscal year 2011," and see. Not, "I'm going to impose my arbitrary aesthetic criteria and invalidate your artistic efforts because your entire venture exists outside my narrow, self-propagating definition of what constitutes good work" and see.

I've actually put off writing about this because, over the last two months, I've been sent to "cover" a couple of ostensibly basketball-centric events that nearly sent me into a rabid, anti-consumerist frenzy.

Example One:

Chris Bosh's Fashion Week Charity something or another a few months ago. What it was about, I still can't fathom, but basically, crowds at Saks Fifth Avenue lined up to shoot Nerf balls at a hastily strung-up hoop with Bosh whilst he was draped in fancy garb. I too tried to "dress up" for the occasion, but as befits my fashion sense (or lack thereof), I looked like an out-of-work accountant from the Great Depression. But lo, there were hundreds of people, forming a serpentine path through the store; all for thirty seconds of…I don't know…a chance to be near a star? To take an iPhone photo of someone famous so you could tweet about it and post it as a Facebook status update? Why, dear Lord, why? All I could think about while staring at the huddled masses was Romero's 1978 film, *Dawn of the Dead*, in which zombies have overtaken a mall and are riding the escalators, numbly staring at stuff they couldn't afford in some half-remembered haze, doomed for all eternity to repeat the pointless, boring, soul-deadening rituals of their former so-called life.

Example Two:

About a month later, like a good little member of the fourth estate/zombie, when beckoned by the powers that be, I too staggered in an undead stupor down to Greenwich Village to report on the unveiling of Carmelo Anthony's new shoe (MJ's brand, natch. That bastid's everywhere!!). I mean...hoo boy...I really almost rediscovered my collegiate, bomb-throwing, radical past on this one. They'd taken over a former gallery and built this monstrous deconstructed basketball court, draped the artistic golem in the new shoes and projected videos of Carmelo discussing his life, game and how his entire ethos has clearly been distilled to a plastic/leather shoe. When the PR guy was showing me around, it took all of my restraint not to ask which sweatshop in China made these items and if Carmelo cared about the deplorable conditions in free trade zones and the worldwide labor struggle in general.

I didn't.

I did ask if the shoe would improve Melo's D, to which the PR flack laughed knowingly. I still don't know if that's a good or a bad thing.

There I was, staring at this unholy creature/creation and all I could think about was the amount of time and effort and thought and MONEY went into trying to sell me or any other potential consumer this product which is, in the end, a shoe.

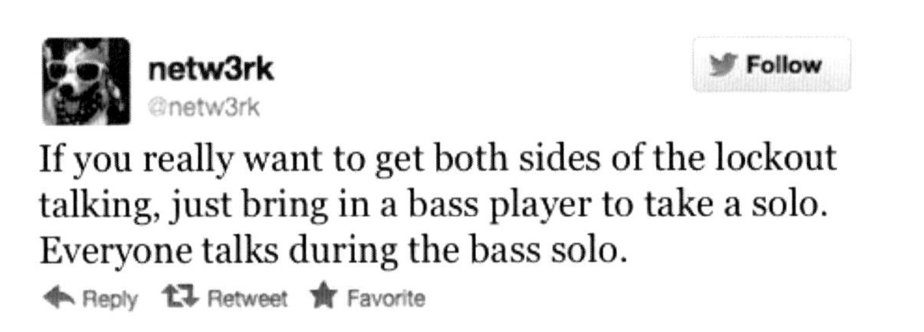

Just a shoe.

In a pique of near-rage, I wanted to scream, "Dear God, what a waste! Look at all the creativity and intelligence on display here, and it's entirely been funneled into selling something that you and I don't really need. Or more specifically, to making you or I feel like our lives are less than meaningful, but if you buy this shoe, well, that will certainly solve all your problems. Get out now! Save yourselves before this beast devours us all!" Again, I didn't do that.

I'm an American. Self-indulgent, holier-than-thou rants notwithstanding, if I'm going to be honest here, I have to admit that I like shiny toys, and I want 'em as much as the next guy and I'll bust my hump to be able to afford 'em. More importantly, I was legitimately excited when I found out I was invited to gawk at Melo's shoe and discuss fashion trends with Chris Bosh. Why? I wanted…something. Fame. Money. Power. Immortality. I don't know. But in my gut, I instinctively knew I wanted or even craved *it*, even if I couldn't begin to articulate what "*it*" exactly was/is.

Alas, one can only confront uncomfortable realities so much of the time. I was talking with my father about the attendant problems with our escapist passion for basketball and he said, "I'm a realist. I've devoted my life to depicting reality."

And I replied (in perhaps, a moment of clarity), "Yeah, but sometimes in order to accurately perceive reality, you need to be able to escape it."

So here we are. The truth is, you watch a ballgame, and you're watching the money. There's no way around that. But another truth is, you watch a ballgame, and you're watching poetry. The pass and fell of these mighty opposites[8] is what makes sport great. It's events like the Melo/Bosh ones and this protracted "negotiation" process that makes one dwell too long on the brutish pole. Or at least it does for me. And I'm sick and tired of following the tweets of reporters hanging out in fancy hotels, wailing like so many undead creatures and waiting for the conclusion of a protracted labor negotiation that in the end is bound to leave both sides royally unsatisfied and in the end, will end up painting everyone involved as short-sighted, money-grubbing vermin.

It's five minutes to doomsday. Owners, players, I implore you. I don't care who gets what. Just give me my game, warts and all, back.

Please?

---

8 Shakespeare.

CHAPTER 2
A KNICKS FAN WALKS INTO A THERAPIST'S OFFICE

BY JARED DUBIN
Hardwood Paroxysm

Parts of this article appeared in Hardwood Paroxysm, *right before the Knicks would face off against the Nuggets for the first time since the Melo trade. At the time, Denver had a better record than New York, and many Knick fans were still upset about the deal. The Melo trade happened, and despite nearly 11 months elapsing, people were still obsessing over it. I thought we fans needed to stop complaining and comparing. The team of my fictional delusion would never be, and it wasn't meant to be. This was our team now, and we needed to embrace it. Eventually, we would.*

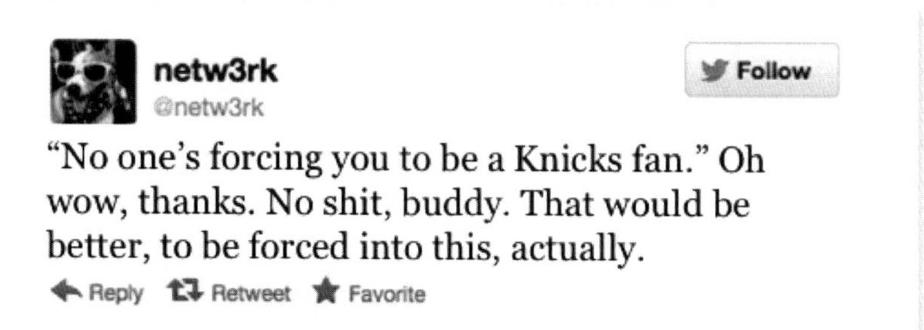

Jared: So, how does this work?

Therapist: How would you like it to work?

Jared: I guess I just, I tell you what I'm thinking?

Therapist: If that's what you want to do.

Jared: I had a dream the other night.

Therapist: And?

Jared: Well, not just the other night. A recurring dream, I guess. Same thing over and over for a while now.

Therapist: Recurring dreams are often a symptom of Post Traumatic Stre...

Jared: Okay, so in the dream, I'm just sitting there on the couch watching the Knicks game...

Therapist: You're dreaming about basketball?

Jared: Well, yeah. I nearly always dream about the Knicks. Anyway, I'm watching the game. And they're losing, like by a lot.

Therapist: I see. And how many times have you had this dream?

Jared: About five times in the last eight or nine days. Anyway, the Knicks are losing. But the weird thing is, it isn't *the Knicks* that I'm watching. They're missing a bunch of players. And they *stink on offense.*

Therapist: Tell me more about that.

Jared: First of all, Gallo's[9] *not on the team anymore*. And that totally sucks, because in real life he's awesome. He's a good shooter, and the Knicks need shooters to space the floor in Mike D'Antoni's system. He can take it to the rim, get some free throws. Gotten a lot better at that over the last few years. He's also a versatile defender, which always helps when your defensive system requires you *to switch on nearly every pick.* So it wouldn't really make sense for him to not be on the team.

Therapist: It sounds like you're quite worked up about this Gallo.

Jared: Yeah. He's one of my favorite Knicks. I really don't understand why he wouldn't be on the team. He's pretty much the perfect fit for D'Antoni's offense, and the crowd really loves him. Oh, also in my dream, Raymond Felton is *gone.* Not so happy about that either, by the way. Dude is a bulldog. He generated some nice chemistry with Amar'e Stoudemire last year, and they're off to a great start to this season too.

In the dream though, Toney Douglas and Mike Bibby are playing a lot of minutes at point guard for the Knicks, and neither of them is very productive at all. And Amar'e just kind of runs into people, throws the ball at the rim and screams, "AND ONE!" like Carlos Boozer. Also he stands around the perimeter and misses jumpers. And lets guys on the other team get open for jumpers. All the time. Basically, Amar'e kind of stinks. That's pretty depressing.

Therapist: Go on.

Jared: Carmelo, though, he's different because Gallo is gone. You know how STAT and the Rooster are playing well together, giving opposing defenses fits at the forward spots because of their ability to hit from outside and get to the rack? Instead Melo is taking lots and lots of shots, and he's making less than he misses, especially from three.

Oh, I should probably back up a bit. See, in my dream Carmelo got *traded* to the Knicks last February. But I know that isn't true, since I remember him signing with the Knicks on the first day of free agency after the lockout ended. He took less money to come to New York to play with his pal Amar'e and everything, just like Donnie Walsh thought he would.

---

9 "Gallo," or "The Rooster," when translated from the original Italian, is the nickname of former Knick small forward Danilo Gallinari.

Not in the dream though. In the dream, Melo came to the Knicks with Chauncey Freakin' Billups, Anthony Carter, Corey Brewer and RENALDO BALKMAN. Yeah, Sayid Jarrah[10] is back on the Knicks. Crazy.

Therapist: And this is different from what you remember?

Jared: Uh, yeah. Everybody knew Melo was *destined* for New York. So why would they have had to trade for him? And they gave up Gallo, Felton, Timo Mozgov AND Wilson Chandler, plus first round picks from now until forever? That's three starters and a serviceable big man. I mean, Chandler wound up leaving as a free agent after the lock-out anyway, so the Knicks could sign Carmelo. But why would the Knicks give up the rest of those guys? Anyway, I'm sitting on the couch watching the game that the Knicks are losing, and all of a sudden my dad walks into the room.

Therapist: Your father came in?

Jared: My dad, yeah. Only, it wasn't actually my dad. In the dream, my dad was Baron Davis. The guy from the Warriors. With a HUGE beard and some weird hipster glasses.

Therapist: Fathers are powerful figures in our lives. We often depend on them when we're most in need. Maybe there's something there? This Baron Davis, is he an important and powerful figure for the Knicks in your dream? Do they *need* him?

Jared: That kinda makes sense. In my dream, they definitely need him. Which is a big part of the problem. Without Felton, they don't have a point guard anymore. They have no rhythm, no flow and no way to get easy shots. You *need* a point guard in order to play Mike D'Antoni's system the way it's supposed to be played. That's why they're so bad on offense. Everybody knows that, especially Knicks management. That's why they hired the poor guy. It's a good thing in real life Raymond Felton's still around to get Amar'e, Gallo, and Carmelo the ball where they need it, because if the Knicks had to depend on *Baron Davis*, I'd probably freak the hell out.

10 Sayid Jarrah is the name of the character portrayed by Naveen Andrews on the surrealistic TV show *Lost*.

Therapist: Why do you call Mike D'Antoni a poor guy?

Jared: Well, in my dream, everyone at The Garden is constantly *chanting* for him to be fired. Lots of boos too. They're even chanting, "We want Woodson!" for Mike Woodson. The coach of the Hawks turned commentator. Except in my dream he's an assistant coach for the Knicks. Actually he's like, the defensive coordinator. He's the guy responsible for all that switching. But the Knicks would never fire Mike D and make *him* the head coach. Not unless the team was in a crisis.

Therapist: It's interesting that you use that word… *crisis*. What else happens in this dream?

Jared: Well, I guess there are a couple of positive things. Tyson Chandler is on the Knicks somehow.[11] The dude who played amazing defense in the finals. In my dreams Dallas let him go, and the Knicks signed him. The best center available, one of the league's best defenders, and the Knicks go after and actually sign him. I know, crazy, right?

And he's playing awesome, actually. Problem is, he's pretty much the only guy on the team who is. But, he's messing up the spacing on offense. Amar'e sucks when he's on the floor with Tyson. There are numbers to prove it and everything.

There's a rookie, too. He's named after a movie. It might have been a TV show too. I forget. The fact that there hasn't been a re-booted, *Love Boat* movie yet continues to astound me.

Therapist: A person that has a movie title for a name?

Jared: Right, but not exactly the same as the movie. He's named *21 Shump Street*.[12] Mike doesn't usually play rookies, so he must be pretty good if he's getting consistent minutes. It seems like he's often playing out of position though. He's not really a point guard. Hmmmm... There's that recurring theme again in the dream, no point guard. Anyway, good on defense, but kind of wildly erratic on offense. Never know what you'll get with that guy.

OH!! Speaking of rookies who actually play for D'Antoni, Landry Fields is pretty *atrocious* in his sophomore season. He's not getting his usual hustle points on offense, he's leaving everyone and their mother open for jumpers on the other end, and he generally looks pretty hesitant with everything he does. And…

Therapist: Jared, I'm going to need you to stop there, because we're almost out of

11 The Knicks acquired Tyson Chandler via a complicated sign-and-trade with the Dallas Mavericks. The Knicks were unlikely to make any big moves this offseason, being well over the cap. But in a shocking turn of events, they amnestied starting point guard Chauncey Billups to create the necessary space under the cap.

12 Of course we are referring to Iman Shumpert. *21 Jump Street* is the name of a 1980s TV police procedural featuring a then-unknown Johnny Depp as well as a recent, re-booted film starring Jonah Hill and Channing Tatum.

time. In the interest of your mental health, I'm just going to tell you something. It may be unsettling at first, and it may make you upset, but in the long run it will help you.

Jared: Uh, okay?

Therapist: You're having what are called *delusions.* You're imagining things that never happened. Living in a world that cannot be.

Jared: That's great news, doc. So the Knicks are fine? Everything's going to be okay. It was all just a dream.

Therapist: No, Jared. The dreams? They're all *real*. That's real life. *THIS* is the dream. I'm *not* real. You're *not* here. The Knicks *traded* for Carmelo.

Jared: Huh?

Therapist: They *gave* up Gallo and Felton and Chandler and Mozgov. First round picks too.

Jared: Are you sure?

Therapist: They *do* stink on offense this year.

Jared: You're lying. With Amar'e and Carmelo and D'Antoni?

Therapist: Garden fans *are* chanting to fire Mike D'Antoni.

Jared: You're telling me "We want Woodson!" actually happened?!?!?! Are you kidding me?

Therapist: They *have* lost five games in the last eight nights.

Jared: Nuh-uh. No way.

Therapist: It all happened. You've just been imagining an alternate universe. But that's not reality. And it's very important that you come to realize that. It's time to snap back to reality, Jared. You need to wake up.

Jared: Oh. Well. Uh, that's pretty depressing.

Therapist: Yeah.

DON'T FORGET ABOUT TONEY DOUGLAS!
(A.K.A. DWTDD[13])

BY SETH ROSENTHAL

Toney Douglas opened the season as New York's starting point guard. The Knicks felt comfortable enough with Douglas as the starter to dispense with Chauncey Billups and bring in the injured Baron Davis, the decrepit Mike Bibby, and the novice Iman Shumpert as their only point guard alternatives to open the season.

At the time, none of the above was considered ideal, but it wasn't regarded as abjectly ridiculous or frightening. In his first two seasons as a Knick, Douglas had rarely performed like a natural point guard, but he was an accurate shooter—best in the league for the post-All Star stretch of the previous season—a competent enough dribbler, and a tireless if sometimes overzealous defender, not to mention a wholesome, likeable fellow. He'd do his best as the starter and the Knicks would get through it.

Only he didn't, and they didn't. Douglas shot and defended miserably from the outset, and the Knicks went 3-4 to open the year before he lost his starting job. From then on, Toney bounced in and out of the rotation, failing to contribute all the while. The popular (and previously positive) refrain "Do What Toney Douglas Do" came to mean, "Miss jumpers, neglect passing lanes, over-commit on defense, and mostly hunker on the bench, coughing occasionally."

There wasn't really a good explanation, either. Rumors seeped out about personal issues, but they never got clearer than just that. Many wondered if the chronic shoulder issue that stemmed from running headlong into too many picks the previous year had persisted through the offseason.

13 This was a footnote explaining what DWTDD means, but we took care of that earlier. So, hey!

Whatever the case, Douglas crashed and didn't recover by season's end. Relatively unqualified guys like Bibby, Shumpert, and the pre-"Linsanity" Jeremy Lin leapfrogged him in the point guard rotation. New York's opening day starter was a Did Not Play, Coach's Decision in four of five playoff games against the Heat, an untrustworthy pariah who'd once been a useful fan favorite. Poor, sweet Toney…

CHAPTER 3
FEAR AND LOATHING IN MADISON SQUARE GARDEN

BY JIM CAVAN
KnickerBlogger.net

(A version of this piece was published in ESPN's TrueHoop blog on February 21, 2012)

On January 20, I learned that I had been granted credentials to cover the forthcoming Knicks-Nuggets game at Madison Square Garden. Now, I'd never felt much like a professional anything, but the prospect of being a working journalist for a night, of covering my favorite team the way many of my favorite writers had for decades, filled me with a pride difficult to describe. The days leading up to the Saturday showdown found me careening between two distinct emotions: elation at checking off one of the boldest bullets from my sports bucket list and terror at how the fast-sliding Knicks would respond to a situation fast spinning out of control. Since the Carmelo Anthony trade nearly a year before, the Nuggets' outwardly ragtag makeup had proven a potent potion of selfless grace and truly gorgeous basketball—everything the Knicks were decidedly not. As such, I was terrified at the prospect of my first game as a reporter turning into a crime scene. It didn't quite end that badly. But it was close.

More than anything, I felt like I was headed to an execution.

Losers of five straight, the Knicks' lockout-shortened season had by late January morphed into a dire downward spiral. Weeks away from the most unlikely of athletic ascensions, hope had given way to Hell in the Garden, and I was going to witness the final descent—a late January home showdown with the Nuggets. Nearly a full year on from one of the most drawn-out and controversial trade operas in recent sports memory, the game had assumed a level of intrigue at odds with what one would expect from your run of the mill, dead-of-winter, cross-conference game; a combination of Denver's endearingly scrappy success[14] and the Knicks' typical train wreck struggles had transformed the meeting into something resembling a capital punishment trial.

Foremost on the gallows was trade centerpiece Carmelo Anthony, whose acquisition had gutted the Knicks of most of their youth and financial flexibility, while bequeathing Gotham a frontcourt tandem redundant in both beauty and moles. If they couldn't get a W, Knicks fans—seldom purveyors of patience or perspective—would likely settle for heads.

I'd attempted more than a few mettle-mauling seven-hour drives into the city over the preceding few months, so this time I decided to drive to New Haven and take the train from there. Though I was coming from New Hampshire, making a 3:50 shuttle—slated to arrive at 45th Street a little after 5:30—seemed a given, freak Nor'easter blizzard be damned. Maybe it had something to do with my "Charles Smith is My Co-pilot"[15] bumper sticker, but a rancid Knicks mojo cloaked me from the go; an accident outside of Worcester on I-495 and the subsequent two-hour, cigarette-heavy delay made me 10 minutes late for the 4:50. I had no choice but to hop the 5:50, which would put me into Grand Central right around tip-off. I bided some time in the New Haven station inhaling a Subway and Gatorade slurry and chatting desperately with Mike Kurylo, wondering over the phone whether being late to pick up my credentials would be grounds for refusal of entrance. How the hell should I know?

The train arrived on time, and I settled into one of the rigid plastic benches for a nervous 90-minute slog. I was definitely going to be late. Question was how late, and could I salvage any kind of decent reportage despite the setback? The train docked, and I hoofed it through the ice-packed Manhattan streets to the MSG media entrance on 8th. Signing in, I found the press pass gatekeeper holding the lone remaining credential aloft, the way a marathon volunteer might a tepid cup of water for the guy bringing in the walkers. I grabbed it, murmuring a barely-audible "thank you" and, brandishing the thing like a paper firearm, blew past half a dozen ushers and security guards until I found myself directly behind the first section of courtside seats.

Mired in apoplectic rage for a majority of the last five hours, I stopped to absorb the moment. I'd been to my fair share of NBA games, but I'd never been so close to the squeaking shoes, the sweat, the voices and movements and flailing limbs. Like a feral child brought to Six Flags,[16] I stood there motionless, mouth agape and mind fast-forwarding through endless hours of 'Bocker brain tape, the gleaming wooden stage flush with long

14 The Nuggets were 29-12 since the Melo trade, at that point.

15 This is a lie, but somebody get on this.

16 Like *Nell* except there's a Six Flags and with a kid instead of a 30-something Jodie Foster.

gone ghosts and faded footprints. In the 90 seconds before security politely asked me to escort myself up to the 300-Level press box, I felt closer to a genuine meditative state than I ever had in any attempt at actual meditation—a moment made all the more surreal by Bill Walker's ridiculous eight-point second quarter barrage.[17]

I trudged up to find the northern press section at this point mostly vacant. Before long, *LoHud.com* blogger extraordinaire Jamie O'Grady sought me out, promising he'd do his best to show a green scribe the ropes. I hadn't been given anything resembling a set assignment for this piece, and truthfully, I hadn't wanted one. I knew it would be extremely difficult to pull back from a fan's natural mechanics and focus on getting a story down, so I didn't even bother. What of it?

Despite playing their fourth game in five nights, the Knicks made a game of what many thought would be a blowout. Still, an ankle and wrist-hobbled Melo struggled, forcing the issue early and often, each ill-advised hoist kicking up the Garden angst another octave. Gallo, meanwhile, was in full payback mode, displaying the kind of versatility that always seemed to exist just beneath his game's surface. Indeed, to someone who questioned the wisdom of the Melo trade as much as anyone, Gallo's performance was equal parts lament-filled and maddening: "Where the hell was this the last three years?" I thought. I'm pretty sure I wasn't alone. Everywhere you looked, jaded fans sported Gallo's once omnipresent No. 8 jersey with equal parts pride and nostalgic protest.[18] Here he was, the 18-year-old kid D'Antoni had once christened the greatest shooter he'd ever seen, assuming a degree of leadership once enjoyed only in instinctual flashes.

Regulation gave way to overtime and overtime to a second, and the Knick offense, with its lack of consistent point guard play by now a fatal flaw, was putting up points almost in spite of itself. Noticeably drained, the Bockers eventually succumbed. Anthony would finish with 25 points on 30 noticeably belabored chucks. The crowd, eager for a scapegoat, had $80 million reasons to target No. 7. In contrast, Gallo, in perhaps the finest overall performance of his young career, poured in a career-high 37 on 19 shots, to go along with 11 rebounds.

The trial was complete; the judgment laid bare in a blizzard of boos. The 20,000-strong jury marched down and out and into the streets and subways, while O'Grady and I headed down to the lower press room, where stale brownies and sandwiches lined the foldout tables, and reporters waited idly for Mike D'Antoni's requisite Q&A. Never one for demonstrability, the Knick skipper, by now feeling the heat beneath his seat, answered the pointed questions, almost all of them relating to Anthony's poor performance, in as measured a manner possible. Ditto the inconsistent defense, the poor shooting, the lack of true

---

17 As it happens, this is also the name of my new post-punk acid bluegrass band.

18 My favorites were two Hasidic Jews sitting side by side, scarfing down tubs of popcorn like it was their last meal on earth.

leadership. What else he could he do?

The dozen or so scribes filed towards the locker room, by then occupied by but two Knicks: Josh Harrellson, whose third quarter wrist injury would result in New York making the most fateful D-League call up in league history two days later,[19] and Carmelo Anthony. Naturally, everyone flocked to the Knicks' supposed savior, seated in a slight hunch facing his locker, slowly peeling pieces of tape from wrist and ankle. For what felt like an hour, Melo said nothing. Finally he stood, lower torso toweled, and cut through the confused mass towards the showers. Apparently, this kind of behavior was rather unusual, as it constituted most of the conversation in the lull before Tyson Chandler and Amar'e Stoudemire's respective arrivals. Never one for curls,[20] my forearms and shoulders ached from holding the voice recorder over the mass of heads attempting to catch the string of humbled offerings—some expected, some candid—ringing forth.

We'll figure it out. I refuse to be part of a losing team. We'll get 'em next time. We just need more time.

Anthony eventually returned to dole out a series of theretofore unspoken confessions: *"Am I shooting too much? Am I doing something wrong? Am I not being a leader?"*

Maybe he's finally getting it, I thought. Looking back, it said more about the state of the team that weekend than perhaps any of us were willing to admit that we'd been reduced to hanging our hopes and prospects on words, and the mere suggestions of change or the possibility of self-reflection, rather than change itself. Nearly four decades removed from the last Knick title, the team seemed perhaps further adrift than it ever had from the ethos that anchored those bygone squads: *Play together, and find the open man. That* was a team. This? This was a collection of 13 guys who, despite playing in a system wholly conducive to those halcyon Holzman principles, were far too often charting their course to one flawed, hazy star.

Afterwards, I met up with O'Grady, Dan Litvin, and Jared Dubin at a nearby bar, where beer and bourbon turned bloggers briefly into aspiring front office sages. *Shumpert's just not a point guard, and I don't even know if he should be starting. If Melo shoots 30 times in a game again, I might move to Brooklyn. I knew it might be rough, but who thought it would be this rough? That's seriously the third Bon Jovi song in the last 15 minutes.*[21]

It wasn't until close to two that I decided to make my way back to Grand Central with the aim of catching the next available train back to New Haven. The night's events still had

19 That D-Leaguer was, of course, Jeremy Lin. Josh Harrellson's injury was one of a string of seemingly innocuous events that had to break exactly the right way in order for Jeremy Lin to explode into the national consciousness.

20 Curls wouldn't help, since it just isolates one muscle. Aspiring reporters should be doing push-ups and pull-ups, which target multiple muscles.

21 No really, it was ridiculous.

me fully wired, so why not use the 90-minute ride back to get some words down, I thought. I knew there was a 2:10 train because I'd looked it up no less than an hour before. Since when does the Internet lie?

When I reached Grand Central's 45th Street entrance, only to find its doors had closed mere minutes before, the day's events hit me like a ton of bricks. A sudden, crippling exhaustion suddenly rendered me incapable of dealing with failure of any kind—least of all the idea that I'd missed the last fucking train home. Scouring the neighborhood for some place, any place to crash, I happened upon a Holiday Inn Express, footed $100 for what would amount to a five-hour stay, trudged up to my room, opened the door and collapsed on the bed. My right heel throbbing after what felt like a fortnight on my feet, I removed my loafers only to find the back of my right sock caked in blood. I took off the sock, tossed it aside, hit the light and in seconds, sank into five hours of dreamless, black sleep.

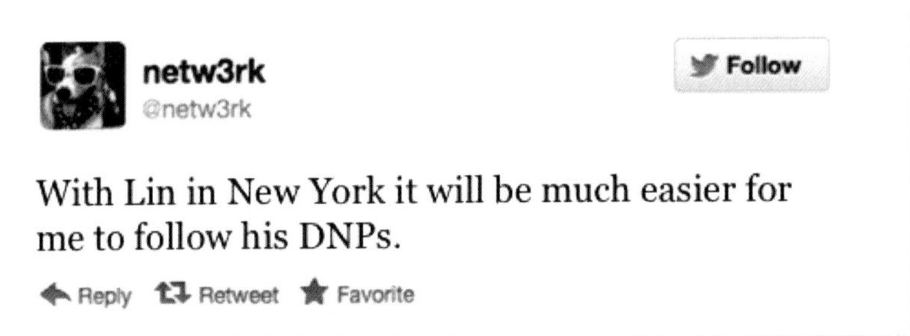

On Sunday morning, showered but hardly clean, I limped back to Grand Central in time to catch the 10:07 back to New Haven. Seated facing backward, we ascended north through the city, past Midtown and Harlem and eventually out of Manhattan. As the train slowly gained speed, that unique feeling of reverse acceleration—a rattling rapture through man's concrete idols—struck me as particularly Knicksian, traveling as I was, faster and faster in what felt like the wrong direction.

Like the Knicks, I sensed the gravity of the city dissipating, its pull and passion fading; my hope for timely arrival had given way to that for mere survival.

Like the Knicks would days later, I just hoped I had bought the right ticket; that the train would take me where I needed to go. Little could we know.

SECOND QUARTER:
FEBRUARY 4 – FEBRUARY 28, 2012

And then, everything changed.

You have to understand that for this fan base, the last twelve years hadn't just been a joyless dirge of downright shoddy, playoff-free (save for a four-game thrashing by Kenyon "Fugazi" Martin and the Nets) basketball—that would have been torment enough. For those either foolishly loyal or loyally foolish enough to continue supporting the denizens of Madison Square Garden, all one hoped for was to be spared a fresh absurdist torment.

The first 23 games of the 2011-12 season, in that respect, were a bit of a respite. Sure, the roster was its usual bloated, top-heavy entity, cobbled together without any actual sense of team building but rather via the capricious whims of (mis)management. But at least this year's model was only a joyless, sullen dirge and not a national joke/endless fodder for the late night comics.

When suddenly, like a bolt from the sky or manna from heaven or a bad, formulaic sports movie, the team's unheralded, unknown, unused Taiwanese American fourth-string Harvard-educated point guard named Jeremy Lin spun the world on its collective ear. Suddenly the Knicks played like a giddy collection of Division III March Madness upstarts, each win more improbable than the next. Last second shots become second nature. NBA icons like Dirk Nowitzki and Kobe Bryant were shredded and turned into true believers and a once moribund season was transformed into the most electric, exciting, national attention-garnering endeavor in recent memory. The fact that the 10-2 streak leading into the All-Star break occurred mostly without the Bockers' two ostensible "stars," Anthony and Stoudemire, did not go unnoticed. To say the Knicks run was unexpected belies the word. To say it was a glorious time does not do it justice. It was as if all the years of sorrow and torment and laughing because it was the only thing keeping one's tears from drowning one's self were being paid back in full. Even though a few nattering nabobs from the chattering class wrung their collective hands about chemistry problems, small sample size and whatnot, for the most part, all was joyous in Knickville.

For the moment…

Amar'e's take on the era: "Phenomenal."

A psychiatrist's diagnosis of Knicks fans: "Psychotic hysteria in response to the deeds of one Jeremy Lin. Patient will be prescribed frequent doses of more Jeremy Lin."

#1 single in Finland: Laura Narhi, "Hetken Tie On Kevyt"

Toney Douglas's Google search history: "Jeremy Lin steroids," "Jeremy Lin cheating"

From the papers: "He won the game for us. He picked us up offensively. We fed off his energy. Down the stretch, we put the ball in his hands. He won the game for us." (Carmelo Anthony on Jeremy Lin, *Newsday*, 2/4/12)

Key stats: From February 4 to March 12, Jeremy Lin averaged 20.4 points per game. In that same span, Carmelo Anthony averaged 16.9 points per game, a full 5.7 points lower than his season average.[22]

---

22 All numberz copyright 2012 by Mike Kurlyo, Numberz Wyzard.

CHAPTER 4
LIN

BY DAN LITVIN
KnicksFan.net

Electricity. It's the one word I'd use to describe Madison Square Garden in the 1990s. There was a buzz about that era and a palpable intensity like an adrenaline storm coursing through the heart of the city. The first Knicks game I attended in person was Game 5 of the 1992 Eastern Conference semifinals, a 94-87 win over the Pistons. Even at a mere 11 years old, I could literally feel that energy pulsing through the building. I've spent the ensuing twenty years vainly searching for that selfsame, overwhelming, and borderline religious fervor.

The Knicks of that era imparted some terrific memories, not the least of which were the '94 and '99 runs to the Finals—both ending in defeat but neither lacking in drama. The following decade, that sensory storm vanished. In its place rose a parched, detached fan base, a zombie hoard aimlessly searching for long-since-rotted playoff flesh.

Things had improved somewhat under the watchful eye of Donnie Walsh, but for the most part the club was still incapable of reviving yesteryear's electricity—"juice," as former Knick GM Ernie Grunfeld used to call it. By February 4, the Knicks' record stood at a disappointing 8-15, and it felt as if the turnaround, begun by Walsh and carried through by newly-anointed GM Glen Grunwald (what can we say, we love us some Gruns), was heading towards the abyss.

It was against this backdrop that the long dormant synapses of unmitigated, intense excitement began to fire again, the idea that the Knicks would become a real team again. And it was against this backdrop that an unheralded, undrafted, Asian American point guard would amplify the Garden's voltage to levels not felt in eons. Up until that Saturday in early February, it had been remarkably easy to identify the lack of a true point guard as the Knicks' chief problem, but with Baron Davis still weeks away from full health, it was difficult to pinpoint a solution. It had become so bad, in fact, that fans had grown angry, desperate, and nostalgic for even the pitiful Anthony Carter.[23] That was saying something.

As the Knick ship steered slowly to a directionless stall, some fans had been prescient enough—or desperate enough, depending on how you look at it—to lobby for seldom used D-Leaguer Jeremy Lin, if for no other reason than it really, *really* couldn't get any worse. An early February *Knicksfan.net* post titled "Knicks Should Use Lin, At Least For Now" posited that Lin might be the one guard on the Knicks who could correct the problem—or a few of the most pressing problems, anyway.

My colleague Jon wrote:

While Lin has a deceptively quick first step he's not especially athletic and he's not a very good perimeter shooter. But what he does have going for him is a nice feel for the pick and roll. He understands how to set up a screen, read how it's being defended, and then at least try to make the right play. He doesn't always succeed but at least he evinces an understanding of what he's supposed to do. On this team that makes him a candidate for some minutes... My guess is that Lin can eventually develop into a competent backup PG, if someone takes the time to develop him.

Thomas Fuller, in *A Pisgah Sight of Palestine* (1650), once wrote, "It is always darkest just before the Day dawneth." For the Knicks, the 2011-12 season had become that very inflection point, that murkiest depth. The Bockers had lost 11 of their last 13 games, a virtual death knell in a lockout-shortened season, where frazzled fans were already trying to process the possibility of falling out of the playoff race entirely. Long before February 4, and true to recent Knicks tradition, fans had commenced the latest iteration of their "Fire [insert name of whatever poor coach happens to be holding the reins here]" chant. This time, the target was Mike D'Antoni, a man who many believed was a false idol, a fraudulent peddler of a bogus system that failed to account for the talents of the team's best players.

The Knicks had already lost to some atrocious squads, including the Toronto Raptors, the Cleveland Cavaliers, and the Charlotte Bobcats, a team that would finish the season with the worst winning percentage in league history. On that fated early February evening, the Knicks hosted the New Jersey Nets and once again found themselves wholly stymied by a seemingly inferior opponent. Completely flummoxed and bereft of other options, D'Antoni—like a poker player holding his last chip on a hand he knows he shouldn't win—turned desperately to the last man on the bench, Jeremy Lin. In the previous game,

23 Anthony Carter, despite being an awful, *awful* basketball player, played a bunch of games for D'Antoni's Knicks in 2011, including all four playoff games. I can't explain why in a footnote. Heck, I could write a whole book on it, but first I'd have to get a degree in psychology.

Lin had seen only six minutes of playing time, missing all three of his shots.[24] Against the Nets, D'Antoni started Shumpert, and inserted Lin with three and a half minutes left in the first quarter, mostly in an effort to give the rookie Shumpert some rest.

And then.... lightning.

Lin played so well that D'Antoni was forced to stay with him. With the score knotted at 82 apiece midway through the fourth quarter, the Knicks were at risk of yet another demoralizing loss to an inferior team. Having already shocked and delighted the crowd to the tune of 16 completely-out-of-nowhere points, Lin would contribute nine more in a 15-4 run, capping off a whimsical, bizarre victory accented by reckless daring and big-moment bravery over 36 captivating minutes. Improbable passes off darting penetration. Blurry drives punctuated by pirouetting finishes. After the buzzer sounded on a 99-92 victory, the fans treated Lin to a standing ovation. Even then, all but a few considered him no more than a curious trinket. I attended the next game against the Utah Jazz. The Knicks were without Amar'e Stoudemire, and then lost Carmelo Anthony to a strained groin a mere six minutes into the action.[25] The euphoric memories are a bit of a blur, but it was clear that something was budding that exceeded easy explication. I was giddy and dazed. As a fan of this long-moribund franchise, I felt like a prisoner released from a dungeon being suddenly reacquainted with daylight.

Soon after, against the Wizards in D.C., Lin came face to face with John Wall, a former first overall pick. The two had paired off once before, in a now somewhat legendary Las Vegas Summer League game that saw Lin outplay the hype-saddled Wall.[26] In their regular season rematch, it was clear that Wall, a point guard who, if nothing else, possesses copious quantities of pure speed, was out for revenge. Decked out in shiny golden shoes, Wall would not again accept embarrassment from someone with such inferior pedigree, finishing with 29 points and six assists. Wall may have "gotten his," but Lin's 23 points and 10 assists spurred the Knicks to a relatively easy victory. Importantly, many of his passes found their way into the waiting hands of Tyson Chandler, a big man whose defensive prowess notwithstanding, wouldn't be mistaken for Patrick Ewing or even Kurt Thomas on the offensive end.[27]

---

24 There were rumors that Lin was receiving playing time only because the team planned to release him after the game against New Jersey and wanted to see how he might perform before he was kicked to the curb. The player the Knicks would have signed had this come to pass? Amityville, N.Y.'s own Mike James, a thirty-something career backup.

25 How depleted were the Knicks at the time? Jared Jeffries and Bill Walker played a combined 63 minutes in that game. Heck, Jerome Jordan saw eight minutes.

26 Jeremy torched Wall in the fourth quarter for 11 points. Wall tried to climb back into his mother's womb.

27 Kurt Thomas can shoot 15-foot jumpers off the pick and pop. If Thomas attempts to score in another fashion, it's best if you cover your eyes. Of course if there are small children nearby, cover their eyes first before your own.

By now Lin's story started to garner national attention. Still, many felt he would come crashing down to earth against the Lakers, the first halfway decent team the Knicks had played in over a week. Lightning could strike once, twice, even three times, but eventually Lin would regress to his D-league genealogy.

Over the previous three games, all Knick victories, Lin had averaged 25.3 points and 58 percent shooting and 8.3 assists. Kobe Bryant, however, was unmoved.[28] Speaking to reporters the night before the Lake Show visited the Garden on February 9, Bryant brushed Lin aside: "I don't really know what's going on too much with him," an irritated Kobe Bean[29] told a reporter. "I don't even know what he's done. Like, I have NO idea what you guys are talking about."

When asked if he would guard Lin if the upstart were to have "one of those games," Bryant broke out in derisive laughter.

As it turns out, Lin delivered the quintessential Linsanity performance. He torched the Lakers for 38 points and seven assists in an effort that *Sports Illustrated* would honor by emblazoning his trademark whirls to the rim on its cover. After the game, Bryant threw roses as only he can. The Associated Press quoted him as saying, "The biggest thing to me is how everybody missed that...They all would be fired if I was owning a team. I hear this stuff about it came out of nowhere, and I think it's [wrong]. You can't play that well and just come out of nowhere. There has to be something there and everybody missed it, so heads would roll."

Having defeated an upper echelon team chock full of bona fide stars, it almost felt like an event horizon had been reached; the fog of doubt regarding both Lin and the Knicks' underachieving season was lifting, to be replaced by an ascendance so immediate and stark that a fear of heights seemed almost inevitable.

---

28 For you ophiophilists that happen to pick this up in a bookstore and are wondering who Kobe Bryant is, he's the guy who gave himself the nickname "Black Mamba."

29 For those that idolize Kobe, this isn't some Kobe-hating inside joke. "Bean" is Kobe's actual, given middle real name. His dad was an NBA player, and his nickname was "Jellybean." Also, Michael Jordan was a million times better than Kobe. It's not even close.

The changing dynamics were all the more remarkable considering that Lin's wingmen in this improbable run were not the injured Melo and Amar'e but castoffs like Jared Jeffries and Steve Novak.[30] As Adrian Wojnarowski, writing for *Yahoo! Sports*, put it: "It doesn't matter what spare parts are thrown together with Lin because he's elevated everyone, transformed five fingers into a fist."

Fans had vilified Jared Jeffries for years for his lack of offense, absent athleticism, and bloated contract. Steve Novak had drifted from the end of the bench in one NBA city to the end of the bench in another, with stops in the D-League in between. Landry Fields' game had been ostensibly neutered by Anthony's arrival but was revived when Lin took over. In fact, the only undisputedly above average NBA player with whom Lin had the benefit of playing during this stretch was Tyson Chandler. These were the players, by and large, that comprised Linsanity. "It's a completely different team," Chandler told reporters. "You can't look at this team the same."

Without its highest-paid stars, teammates were forced to trust one another and their coach. In the span of a few weeks, Lin had taken this team of misfits and made them one of the most feared and unpredictable teams in the league.

I got my second taste of live Linsanity on February 19, when the Knicks took on defending champion Dallas Mavericks at MSG. Before the game, I'd turned down an opportunity to sell my two $140 tickets for $1300. The answer is, "Yes, it was worth it." Lin again led the Knicks on the comeback trail, nailing a huge momentum three pointer over all seven feet of Dirk Nowitzki in the fourth quarter en route to an awe-inspiring performance.

"He looks a little bit like Steve Nash out there," Jason Kidd would remark after.

"I think they found something in Lin," Jason Terry chimed in.

---

30 In 2011, Jared Jeffries was released by Houston, and Steve Novak was released by both Dallas and San Antonio. The pair managed to "meh" all three Texas teams in a single year. Yes, I just used "meh"as a verb.

Observers, many of whom had been cautious about Lin's prospects, began to agree not only with the found money aspect of the narrative but also with the loftier comparisons to Nash and other elite point guards. Folks who derived statistical models for a living began predicting what Lin would ultimately become. After studying Lin's performance, *ESPN.com*'s John Hollinger concluded that his closest statistical comparators were Isiah Thomas, Kevin Johnson, and Russell Westbrook—a none-too-shabby collection of contemporaries. Carl Bialik of the *Wall Street Journal* undertook a similar exercise and found favorable comparisons to John Stockton, Chris Paul, and Steve Nash. Bradford Doolittle of *Basketball Prospectus* wrote for *ESPN.com* that across a full season, only Chauncey Billups, Penny Hardaway, LeBron James, Kevin Johnson, Magic Johnson, Steve Nash, Chris Paul, Mark Price, and Deron Williams had achieved, like Lin, a "23.8 percentage usage rate, .594 true shooting percentage [31]and an 8.7 percent assist rate."

Whether or not these projections would come to fruition over the course of a season or a career remained the subject of much debate. What was not up for debate was that we were witnessing moments magical enough to give rise to such comparisons in the first place. It was just about as exciting a phenomenon as you could hope to witness in the world of sports. This was manna from heaven. It was fate smiling on a wayward, lost franchise and their fans. After years of terrifyingly awful management and organization folly bordering on the ridiculous, the Knick faithful deserved it.

When Anthony returned from his groin injury, D'Antoni's vindication quickly gave way to Anthony's vindictiveness. The Knicks would lose six in a row and eight of their next ten, returning them to where they had started before Linsanity. Anthony in particular appeared unwilling to buy into many of the principles displayed by Team Linsanity—spacing, movement, and passing, the hallmarks of Mike D'Antoni basketball. Melo could often be found sulking on the bench, declining to participate in huddles, lashing out at teammates for failing to bend to his will when he broke plays, and perhaps most damningly, frequently outhustled.

Ultimately, Linsanity ended just as quickly as it began. But nothing can take away the three weeks of stunned, beautiful bemusement, and the countless puns—LINCREDIBILE; KNICKS DO NUMBER ON JAZZ; LINNING STREAK; MAY THE BEST MAN LIN; LINCREDIBLE JEREMY WIN!; LIN & BEAR IT!; vaLINtine's

31 True shooting percentage is a measure of how well someone shoots. It's superior to field goal percentage because it accounts for three pointers and free throws.

GIFT; AMASIAN;[32] DISH UPON A STAR; LIN-PHOMANIA; WINSANITY—that made the rounds on pixels and pulp alike.

When a repeatedly discarded and underestimated player—days away from being cut and losing grasp on his NBA dream forever—was given one last chance, his only chance, he made the most of it. He was named NBA Player of the Week. He averaged more points in his first six starts than any player in the history of the league. He was the first player ever to record at least 20 points and seven assists in each of his first five starts. From February 4 to 19, Lin averaged 25 points per game on 50 percent shooting, dished out 9.2 assists, made 2.2 steals, and snared 3.7 rebounds.

As Will Leitch wrote for *NYMag.com*, Lin's emergence transformed the Knicks from a "group of cranky malcontents" to a "giddy batch of hopping, laughing lunatics in the span of about 72 hours."

However fleeting, he did the same for the fans. He turned the power back on at MSG. Electricity permeated it again, and in turn it permeated us.

DON'T FORGET ABOUT JOSH HARRELLSON (A.K.A. JORTS)!

BY SETH ROSENTHAL

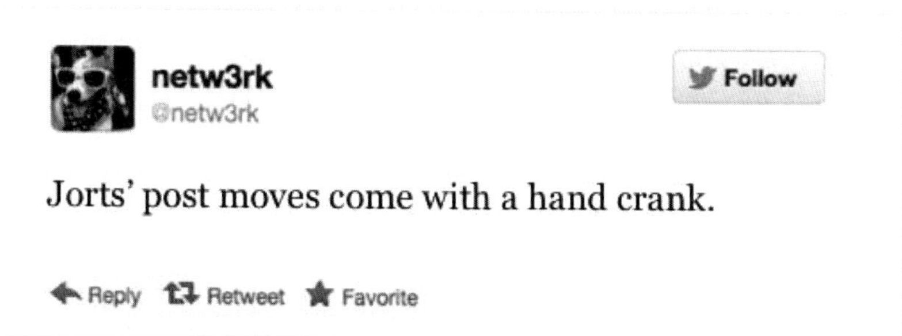

The Knicks went out of their way to get Jorts. New York didn't own a second

32 Some of the pun-tastic headlines that Lin inspired veered towards offensive Asian stereotypes. During the Laker game, certain fans could be seen holding up signs referring to Lin as the "Yellow Mamba." An *ESPN.com* producer was fired for posting the headline, "A Chink in the Armor," at the conclusion of the Knicks' seven-game winning streak.

round pick in 2011, but paid cash to the New Orleans Hornets for the privilege of drafting Josh Harrellson, the big, sweaty bumpkin wedged into John Calipari's otherwise glamorous squad of surefire prospects. It seemed an inspired stroke of draft night whimsy that New York, of all locales, would make a deliberate move to welcome in a Missouri-bred, Kentucky-branded, buck-huntin', catfish-noodlin' late-bloomer most famous for wearing denim shorts (the genesis of the term Jorts) to a recruiting visit and pegging Jared Sullinger with a basketball.

Knicks fans without much knowledge of "Jorts" got to googling and discovered a 6'10" tall, 5'10" wide (educated guess) white dude, a giant grinning hallux wearing a taut basketball jersey and a gel-embalmed widow's peak. He was presumptively assigned the role of "gritty bench thug," and mostly disregarded during the lockout, with the exception of the night in August when he reportedly thwarted an aspiring drunk driver in a Lexington parking lot.

For the most part, Harrellson embraced and fulfilled that assignment. He earned sparse bench minutes from the outset, establishing himself as a willing rebounder and wanton fouler with a mild case of the jitters on offense. Jorts promptly found his legs on the hardwood, though, proving with remarkably nimble footwork and apt timing that his "dirt strong" game—Mike D'Antoni's words—was more of an exact science than it seemed. Jorts revealed an unexpected degree of finesse ("revealed" by unwrapping layers of bacon, of course), chiefly with his remarkable outside prowess. He'd launched a couple threes to open the season, but really let loose on New Year's Eve, when Amar'e Stoudemire sat with a sprained ankle and New York needed a frontcourt boost. Jorts hit four threes in that win over the Kings, then went 12-34 from downtown through the first three weeks of January, attempting the last of those long bombs with a freshly broken wrist.

Despite Harrellson's apparent disregard for his wrist fracture, the injury kept him out until early March. He returned in the middle of the post-Linsanity swoon and just days before D'Antoni fled town. From that point forward, Jorts's minutes were spotty and spent mostly inside the arc, with just four more made threes leaving his paws the rest of the season. "Gritty bench thug" isn't a bad descriptor of the niche Harrellson eventually found as a Knick, but it does belie the surprising degree of polish the guy demonstrated at times. There was deftness in those Jorts.

CHAPTER 5
THE ALGONQUIN ROUND BALL
FEATURING JARED DUBIN, NETW3RK, AND SETH ROSENTHAL

Unless you'd been living under a rock—and even then I still think you would have heard about this—you knew there were some big things brewing in New York at the end of February. Jeremy Lin had emerged from out of nowhere and taken the nation, the city and the fan base by storm. One of the best parts about this story is that everyone—bloggers, fans, and basketball atheists alike—really did experience it together. With that in mind, three Knicks bloggers took to emailing back and forth about their experiences coming to grips with Linsanity—individually, together as Knicks fans, and as part of the larger NBA universe. Portions of this chapter originally appeared in Hardwood Paroxysm *in late February 2012.*

Dubin: We're now through Game 11 with Jeremy Lin as a regular presence in our lives, and it undoubtedly feels different today than it did when he first surfaced on February 4 against the Nets. He's gone from one-hit wonder to improbable surprise story to international sensation in a span of just over two weeks. Everyone's experiencing Linsanity in some way, but as Knicks fans, we're really right in the middle of it.

The ride has been incredible. As I've already said, it's often left me speechless. Sometimes because I have nothing to say, and other times because I don't want to say anything for fear that I'll somehow mess things up. That's the nasty undercurrent with Knicks fans; we're so neurotic, so used to everything going so spectacularly wrong that even when we're in the middle of something as joyous and undeniably terrific as Linsanity we still have that creeping doubt in our minds. The one saying, "This will all end soon. It always does." We're always waiting for the other shoe to drop.[33]

When I see a short-handed team fight for a huge comeback win against the Mavericks, it feels like that apprehension is subsiding, and that I can trust this team. But then they come out and lose to the Nets, and we're right back where we started emotionally. I'm confident in the kid and I'll defend him against anyone who says he can't play or he's not that good, but sometimes I can't help but feel like this all isn't really real.[34]

So my question(s) for you guys is (are): how have you felt about your experience with Linsanity? Do you feel that doubt at times, too? Have you at all worried, despite

33 It does.

34 It's not.

everything we've seen and read and written about him that points toward any of this being sustainable, this will all come crashing down? Has that feeling waned? Have I asked enough questions yet?

Rosenthal: Pretty much the whole time I've been waiting for the other shoe to drop or whatever, but recently, I've found myself able to take my eyes away from him and not fretting over every little move he makes. I trust him and, I'm scared to admit, I must be starting to feel like he's reliable. I know this because I looked at his stat line after the Atlanta game: a 17-point, nine-assist outing that, if it came from a Knick point guard in January, would have made me soil myself. And, you know, I didn't soil myself. I must have involuntarily adopted a set of expectations for Jeremy.

But then, as a self-centered Knick fan, I'm now making myself worried that I'm taking the kid for granted and he's suddenly going to fail or melt into a puddle or something. The Knicks can't just establish something good and keep it that way. That opposes everything I've been conditioned to expect from this organization. Oh god, where's Jeremy? Has anybody seen him recently? Did he disappear? Are we even sure that he ever truly existed?

See what you did, Jared!?

netw3rk: Lin's emergence has been, for me, the most purely enjoyable stretch of Knicks games I've experienced. Even when the Knicks were contending in the 90s, the pressure of trying to overcome Chicago or fend off Indiana and Miami was suffocating. The Knicks were truly the bad guys in black sneakers and shaved heads then, an eminently hate-able (if you weren't a fan) team who physically pulverized opponents. The winning was fantastic but it was gone about with an air of grim duty and a hint of maliciousness. The incredible run during the '99 postseason comes close—the peaks were higher—but there wasn't the sustained feeling of happiness game in, game out.

What links '99 and Lin's emergence, in my mind, is that both were near total surprises. I was pretty darn sure New York's window was tightly shut in '99, at least as far getting to the finals was concerned. I was also pretty sure Lin—despite flashes—didn't have anything close to the games he's put up in him. I still find myself jolted by an electric bolt of surprise when he makes a layup.

Now, 20 points and eight assists is not sustainable. I'd be happy with 12 and 6.5. I am concerned about the effects of the hype, and the hype backlash, on him. In general I think I feel like most Knick fans do towards him: protective.

Dubin: I definitely feel you on the "protective" bit. Especially after the game against the Heat, when all the, "Oh, man. Linsanity is OVER. This kid is done and it was alllllllll a fluke," type of people came out, I felt the need and the urge to rush to his defense. Lin got so big so fast that there are people who are now rooting for him to fail because of it; and at times I really do feel like it's on us as Knicks fans to be, for lack of a better description, his offensive linemen, and protect him from all that criticism. I think that plays into Seth's point about developing a set of expectations for him too. I think a lot of people did, and that made it pretty jarring to see him fail so spectacularly against the Heat, even if we should have expected him to struggle, though not to the degree that he did, given their defensive strategy and the athletes they can put on the floor.

You bring up a good point about the 90s Knicks as well. Those teams were flat out awesome, and I loved rooting for them. They're the teams I grew up on and I'll remember them for the rest of my life, but they were not fun teams.[35] Fun to root for, sure, but not to watch play the game of basketball. It was dirty, it was ugly, it was hang 'em and bang 'em, knock-down, drag-out basketball, and not a soul would tell you it was aesthetically pleasing.

But I've had so much goddamn fun just watching Jeremy Lin and these Knicks play basketball. The fact that I share the experience nightly with everyone on Twitter, in my writing and in reading everything that everybody else writes probably plays into that, but he's also just been exhilarating to watch—the last-second heroics, the charisma, everything. Really, the Knicks have never been this fun.

How has all that played into your experience with Linsanity? Do you feel that Twitter and writing and just generally being part of the NBA blogosphere has elevated it to another level than it would be at if you weren't an active participant in those arenas? I know that sometimes I feel like everyone is watching us very carefully as Knicks fans to see how we react to all this. Some people are waiting to jump down our throats for getting too high or too low on the kid, others are just watching for entertainment purposes, but it does feel like there are eyes on us for sure. Do you feel that way, too?

netw3rk: Twitter has definitely magnified the Lin experience for me, mainly because Twitter has become the de facto way I talk about basketball. I think people are watching Knicks' fans reaction to Lin for a few reasons, the largest one being: The Knicks are viewed as a star-crossed franchise, so something unexpected *and* good happening is a new experience for everyone involved. Various flavors of Knicks/New York/Big Market hate manifested in a hope that Lin fails so that people can rub it in our faces.

Dubin: That Knicks/New York/Big Market hate hasn't really hit Lin yet, but it has to be coming pretty soon, right? You just don't have an extended period of success in New York without people turning on you unless your name is Derek Jeter. Right now, there's the occasional salt thrower on Lin's game, but nothing too crazy. I could really see him engendering big time haters because of how big his story has gotten and how fast. Do

35 Asked to spell "fun," Charles Oakley responded "F-U-C-K Y-O-U." Wikipedia that. I dare you. Actually don't. It's a lie. Unless we all add it to Oak's Wikipedia entry. Then it's true.

you guys see that in his future?

Rosenthal: I think a big part of Lin's avoiding the "HE'S FROM NEW YORK! BURN HIM!" backlash is how consistently he deflects even remotely seedy attention. Who knows how he'll fare over the long run, but guys like Jeter have gotten in trouble in part because they occasionally get caught with their hands in the cookie jar, gallivanting around with some hottie, or living lavishly in some other realm. Lin's been pretty outspoken in saying he doesn't want to become like that, though there's no way of knowing how he might change as he gets comfortable here. That isn't to say that a guy having a nice car or keeping hotties around as company is a bad thing. People just flip out about those things.

I think that, because it still seems so improbable and because it's New York, people will be especially prone to dissecting Jeremy's game as long as he stays reasonably good. As long as his apparent flaws are confined to the basketball court, though, I can't see it getting out of hand the way it does for some other New York sports figures.

Dubin: That's a good point you bring up about how Lin deflects that type of attention, because I feel like he tried to do that with all the on-court attention when this whole thing first started as well. He would constantly bring up guys like Jared Jeffries and Shump and Tyson in interviews as if to deflect praise away from his own efforts and highlight theirs. He's seemingly made a lot of the guys on the team more confident in their abilities through his confidence in them on the court and also the way he speaks about them off it. In terms of his on court performance, though, do you feel that people will be quicker to pounce on him if he does wind up falling back to earth because of the way this story developed? They almost have to, right?

Rosenthal: Oh, definitely. It's bound to happen, and it's already started to happen after a bad game or two. I just hope people are directing some of that grouchiness toward the entities that blew so much hot air into Lin's story (none of whom were Jeremy himself).

netw3rk: I'd probably pounce on him if the situation was the same except he played for someone else.

Dubin: It feels like the All-Star break may help some of the hysteria around Lin die down, especially since the Knicks only play once between now and Sunday. The fact that it was a continually ongoing story that had so much momentum because it felt like the Knicks had a game every day played a huge part in Lin's rise.

So, where do you see this all going from here? Will the crazy amount of hype die down naturally or will it take a string of bad performances like the one against the Heat last week? As a Knick fan, would you like it if the crazy hype and all the pomp and circumstance surrounding this situation all just went away? I know that, when I saw how exhausted he looked during All-Star Weekend, I definitely wished a few times that he could just play basketball. I'm happy for the kid that he's a big story and he deserves everything he gets, but I can't help but worry that it'll burn him out (which again, plays into that creeping doubt that all Knicks fans experience).

Rosenthal: I do think the All-Star break and the loss to the Heat and just the passage of time and people's wandering attention will allow the "Linsanity" stuff to die down somewhat. And he's bound to have some more bad games here and there, which will throw off some of the more casual observers who don't really understand what's going on and expect perfection.

In the long run, I think everybody settling down can only help matters because, like you said, he's going to get burned out if he remains such a point of focus off the court. I just hope that it doesn't turn into a case where, instead of just retreating from the story, folks feel the need for backlash. It's not Jeremy's fault that he's been lionized and I hope he doesn't have to bear the brunt of any unreasonable outrage if he makes mistakes that are perfectly acceptable from an undrafted second-year player.

netw3rk: I don't think anyone knows where Lin's ceiling will end up. But if I had a gun to my head, right after wetting my pants, I'd say Lin will develop into a better Ramon Sessions, adept at the pick and roll, not a great shooter.

The upside to Lin's well-publicized weaknesses (going left, shooting) is that these are things that should improve over the course of his career. He's only 23, after all. Knicks fans need to prepare themselves for more games in the mold of the Miami game as defenses catch up to him. He's not going to fix his problems this season.

So, get ready for some backlash.

CHAPTER 6
LINSANITY, INDEED

BY ROBERT SILVERMAN
KnickerBlogger.Net

After two games, the unimaginable occurred. In a season that looked like it would exemplify Michael Ray Richardson's famous epithet, "The ship be sinkin'," out of nowhere, the Knicks and their fans were completely overwhelmed by a giddy sense of probability-smashing enthusiasm unlike any seen in these quarters for decades. After his first two starts, in which Jeremy Lin seemed to be channeling a combination of Steve Nash, Allen Iverson and a mythical, multi-armed, animal-headed, omniscient Zoroastrian deity, Mr. Silverman penned the following.

I've got a story for you.

What if I told you the Knicks would be without Amar'e Stoudemire, Carmelo Anthony, Baron Davis, and for most of the night, Tyson Chandler, and they'd win?

What if I told you their leading scorers would include Steve Novak, Jared Jeffries and Bill Walker, and they'd win?

What if I told you the catalyst for this seemingly impossible concoction would be an undrafted, Taiwanese American, second-year point guard from *Harvard,* of all places, who had spent the weekend crashing on his teammate's couch and possessed a total of zero starts and 39 career games played prior to the game versus New Jersey?

What if I told you that on Monday, this international man of mystery would net 28 points on 10-17 shooting and hand out eight assists to lead the Bockers to consecutive wins for the first time since January while also miraculously being thrust forward and made the centerpiece of the most recent ad campaign concocted by MSG's Don Drapers?[36]

It may sound like I'm pitching a fairly tired, hackneyed premise for yet another "rags to riches" sports movie, but luckily for the Bockers, the above isn't my take on a hoops-centric remake of *Damn Yankees*;[37] it actually happened. Out of nowhere, the team's

36 Don Draper is the name of the lead in the AMC television show *Mad Men*, which depicts the trials and travails of a group of advertising executives living in New York City during the early to mid-1960s.

37 *Damn Yankees* (1956) is a musical comedy in which Joe Hardy, the hero, sells his soul to the Devil in order to become a world class baseball player. He then joins the perennially moribund Washington Senators in the hopes of defeating his and the Senators' longtime tormentor, the New York Yankees.

season-long point guard issues actually appear to have been solved by one Jeremy Lin, who, as recently as a month ago, was toiling in obscurity for the Erie BayHawks of the NBA Development League.

Totally implausible, am I right? Makes no sense at all. Well, I'm just as gobsmacked as you are. In fact, the phenomenon that is Jeremy Lin has been such a revelation to Knick Nation that it almost defies describing. We are truly in a state of ecstatic glee. But this is more than just discovering a hidden talent, winning games, improbable victories or even the dramatic 180-degree turn of yet another season that seemed to be careening towards a grim conclusion. No, this is something else. I believe there are a couple of things at play here; one is secular and the other is decidedly not.

As to the former, there's a reason that Hollywood has essentially been recycling the same plot lines for the past century or so.[38] It's because in many (or even most) facets of life, the good guys don't win and David doesn't sling the rock that bops Goliath. Most people don't even get the *chance* to play David or Goliath. Far more often than not, rather than an epic, Manichean struggle, one's life is filled with half-measures, unsolvable or inscrutable problems, maddening inconsistencies and vast shades of grey.

So when a player like Jeremy Lin bursts on to the scene, each foray to the rim a spectacle of effort and improbability, and actually vanquishes seemingly far more powerful foes while bolstered only by a skeleton crew of the Knicks' lesser (to say the least) lights, you can't help but think it's fiction. More importantly, because his tale is so utterly improbable, you want nothing more than to identify, to feel as though his victory is truly ours. It's as if, like the old yarn about Flo Ziegfeld and the chorus girl,[39] when the moment struck, we were able to act with bravado, perform Herculean feats, overcome a lifetime of timidity, rise from obscurity and become the city's latest darling.

This may read like yet another example of a New Yorker's bias or sense of superiority, but gushing adoration of the underdog is practically a requirement for citizenship in Gotham. In this instance, we're allowed to crow a little. By any statistical or anecdotal measure, making it in this town is darned tough. So when an R.A. Dickey[40] or a Victor Cruz[41] actually *does* overcome the overwhelming odds and vanquish a far mightier foe,

---

38 To be accurate, they're still going by the rules laid out in the year 335 BCE in Aristotle's *Poetics*.

39 Florenz Ziegfeld was a producer of theatrical spectaculars at the turn of the century called *The Ziegfeld Follies*,which featured a large chorus of dancing girls. As the story goes, one night, the star of the show was unable to perform so Mr. Ziegfeld chose an unknown chorus girl to take her place. She excelled, instantly transforming her into a star.

40 R.A. Dickey is an All-Star pitcher for the New York Mets. He was an unheralded player who resuscitated his career at age 37 by developing a knuckleball to become an unexpected star. Alas, the Mets were still awful.

41 Victor Cruz is an All-Pro wide receiver for the New York Giants. He was an unheralded, undrafted player who became one of the leading lights in the Giants' run to the Super Bowl, whereas Lin did not achieve the same degree of success, to say the least.

it's a straight shot to the heartstrings of every New Yorker trying to scrap and claw his or her way to the top. Not that one isn't thrilled to death when Carmelo or Amar'e drains a game-winning shot, but it just doesn't have the same *oomph* as when Jeremy Lin nails an off-balance three with the shot clock running down. Why? Well, Melo and STAT are Stars with a capital S.[42] Playing the hero is what they're expected to do. While a child can and will closely identify with the league's great players, as an adult, our limitations seem fairly set in stone, and seeing Melo's or STAT's or LeBron's exploits as a paradigmatic of our own mundane daily endeavors becomes a far more daunting proposition.

Truly, the last cager[43] I can recall who made the great, huddled masses rise as one the way Lin does was Latrell Sprewell, and, once again, as has been the case throughout the team's tumultuous history, a redemptive arc was at the core of the story. The overwhelming adoration that was showered upon Latrell as he powered the 1998-99 Knicks to an improbable title run[44] was a classically New York embrace of a figure who had been a league-wide pariah the year before he came to the big city.[45] And just like Latrell, Lin is a creature born out of the fulfillment of an epic, Joseph Campbell-esque narrative. He wasn't born to be a star. He's the archetypal little guy tryin' to make it big; the direct descendant of Charlie Chaplin's Little Tramp. While the Frazier-DeBusschere-Reed-Bradley crew may reign in a pantheon reserved for exalted immortals and Patrick Ewing will be forever cast as a noble, everlasting warrior, our hearts hold a special place for John Starks—a cat who was bagging groceries in Tulsa not long before he posterized Pippen and Grant with a replica of one of Michael Jordan's airborne efforts at the end of game two of the 1993 playoffs and Anthony Mason, an undersized, point power forward who was summarily released by the Nets and Nuggets and even some team in Turkey before emerging as the scowling, menacing, brutish, hairstyle-as-a-billboard-for-Dadaist-prose heart of Riley's not-quite-great 90s teams.

In most instances, Lin's eruption would merely be an example of yet another athlete, like Dickey and Cruz, who pinged an archetype near and dear to denizens of the

---

42 Whether or not Carmelo Anthony and/or Amar'e Stoudemire should be called "stars" is another, far longer thought entirely.

43 In the early days of professional basketball, the newspapers would colloquially refer to the players as "cagers." The name arose because the local gyms where the games occurred often featured a mesh cage around the court to prevent the fans from hurling objects at the opposing team. It's not unlike the scene in the movie, *The Blues Brothers*, when the band plays a gig in a Country/Western bar and the saxophonist dryly notes, "Chicken wire?"

44 The Knicks became the first eighth seed in NBA history to reach the finals, upsetting the heavily favored Miami Heat and Indiana Pacers in the playoffs, following a lackluster regular season in which they finished with a 27-23 record.

45 Latrell Sprewell was suspended without pay for the remainder of the 1997-98 season for choking his coach, P.J. Carlesimo during a heated argument during practice in which P.J. purportedly told Latrell to "put a little mustard" on a lackadaisical pass.

Big Apple. In the case of this particular squad and this particular player, you cannot understand the degree of exultation without taking into account dredging the bottomless sea of horrors that the die-hards have been forced to endure over the last decade. If you haven't been following the New York professional basketball team closely, it's not just that they've produced a series of execrable teams. They've humiliated themselves both on and off the court. They've literally been a collective punch line, a bad practical joke, cautionary tale, and the exemplar of everything that is wrong-headed, vicious, and stupid in professional sports. Within the context of a valley of futility so deep and wide, it's not surprising that this wholly unexpected peak named Jeremy Lin would inspire a reaction that most closely resembles…well…religious ecstasy.

That may seem like an extreme comparison, especially in reaction to someone who has only started two games. But bedraggled Knickerbackers have spent so long wandering in a desert of absolute despair that that the glimpse of an oasis that Lin has provided will invariably be lunged at with a rapturous-crazed frenzy. To ask us now, after what seems like an interminable period of hopelessness, to maintain some measure of distance would be an act of pure folly. Heck, when one is as desperate as we have been, it's nigh impossible to act in a rational, objective manner.

Then again, that's the nature of the beast. The words "rational behavior" and "sports fan" do not belong in the same conversation. The love that one has for a team or a particular player on said team never was and never shall be a *rational* endeavor. Ask your standard-brand long-time Red Sox or Cubs devotee why they've remained in the fold after the decades of bone-crushing defeats and constant, casual anguish that said devotion has wrought upon their souls. There is no *logical* response. Fandom is, by definition, an act of faith: a set of traditions, values, icons, and fables that are passed down by father/mother to son/daughter from generation to generation. We bear our scars with a measure of pride. We spend hours scouring box scores and devouring information, arrange our social lives around a three-hour televised event and base our entire emotional well-being on the outcome of a three-hour contest in which a group of oversized individuals scamper about in the undergarments because, in the end, it's a form of *worship*. We are *fanatics*.

From the outside, it sure might resemble madness. And when the team/altar to which one has chosen to serve is as preternaturally doomed as the Knicks are, have and probably forever will be, any human with a reasonably functioning pre-frontal cortex who used every inch of his or her gray matter to assess pros and cons of the aforementioned equation would walk, nay, run away, possibly screaming and ripping one's hair out by the roots.

The reason we members of the tribe behave in such a manner is because we are oh-so-acutely aware that we have absolutely no way to impact whether our team will win or lose. We can adore the Knicks or any other team to the nth degree, but the strength of our passion won't make a single shot go in or a ball bounce the right way. That utter loss of control, of our ability to shape the world (or at least something as central to our lives as a sporting contest) according to our will, frankly, is terrifying. So what do we do? We do the same thing humans have been doing since time immemorial when confronted with the realization that we're powerless, that things happen both for good and for ill for no discernible rhyme or reason, and forces far greater than we'll ever truly understand decide our ultimate fate. We pray or superstitiously wear a lucky shirt or eat the same meal or whack ourselves on the forehead with a ball peen hammer or whatever ritualistic or spiritual appeal might allow us to cling to the illusion that in our ministrations we might have some tiny influence on events of great import like a basketball game.

That's not to say that we fans are all closeted martyrs. We suffer the daily torments of fandom because there exists the slim chance that one will get to experience a borderline-ecstatic state of being that comes with watching Jeremy Lin play. We might even say with a straight face that the eternal devotion of our soul is a fair price to pay in return for the Universe or God or the fates or whatever bestowing upon us a franchise-altering savior, dropped straight from the heavens (or at least the rafters of Madison Square Garden). That's the contract that we've entered into. Because of Lin, our faith and our fandom (for the moment) are justified. We can point to him and say, "See! Prayers DO get answered. There are such things as magic and beauty and maybe even the occasional divine intervention in the world. This is why we accept the agony and allow ourselves to be subjugated to a capricious, ephemeral entity that may or may not love us or even care that we exist![46] Jeremy Lin renders your logical assessment of our fanaticism moot! MOOT, I SAY!"

That's where we are right now; riding the crest of the wave, hoping against hope that it doesn't break. And, even if this minor miracle does collapse upon itself, if Lin turns out to be yet another false prophet, at least we can say that, if only for a fleeting moment, the joy and the love we felt was real. I hope and pray that it's only just begun.

---

46 Just to be clear, said entity is the Knicks and/or James Dolan, not the Almighty.

DON'T FORGET ABOUT BILL WALKER (A.K.A. BILLY)!

BY SETH ROSENTHAL

Bill Walker didn't do too much. He opened the year as a regular backup wing and saw even more minutes when Carmelo Anthony went down with a wrist injury. He'd use those minutes, though, to commit some fouls, maybe hit a couple threes, smirk a lot, wear a headband, then head back to the bench (where he'd continue to smirk and wear a headband). Even when given ample minutes, Walker's play rarely inspired.

Rarely, but not never. On January 27, the Knicks were in Miami, home of the eventual NBA champion Heat, without Carmelo Anthony. They didn't have a prayer of winning, and they seemed to know that from the outset. As a team, New York tossed up more threes (43) than twos (41), and hit a fairly impressive 18 of those outside shots. Leading the way, of course, was our friend Bill, who seemed to have Coach D'Antoni's consent to heave with abandon. Walker hit seven of ten threes that night, including a delirious nine-point run in a minute-long stretch of the third quarter that included a blind, turn-around, fall-away three from the elbow and a shameless bank shot from the top of the arc.

New York lost by ten, but they totally would've lost by 30 if it weren't for Walker. And that was the Bill Walker game.

CHAPTER 7
NOTES FROM A KNICK JOURNALIST UNDERGROUND

BY JAMIE O'GRADY
The Journal News

There are but a few times in one's life–usually with the wind at your back and luck smiling upon you–when a crossroads of the improbable and the inexplicably possible reveals itself. For many, disbelief and paralysis preclude a deviation toward exploration. For Mr. O'Grady, however, an opportunity not taken has never been an opportunity at all, so when he was given the chance to cover the Knicks for a local newspaper, he slammed his foot on the accelerator and never looked back.

December 25, 2011: BOS @ NYK

Never in my entire life have I felt this small. And not entirely in the physical sense, either. After all, at 5-10, I am hardly the smallest guy in the room. (No, that honor goes to the legendary, yet diminutive *Daily News* scribe Mike Lupica.) This is something different altogether. This is Madison Square Garden. These are the New York Knicks. And here I am masquerading as a legitimate sports journalist with zero formal training and a recurring tendency to color my "reports" with just enough wit to entertain and not nearly enough impartiality to be taken seriously. Not that it matters, ultimately. I am in the room, so to speak, and for the select few of us about to bear witness to what will arguably become the wackiest single season in professional sports history, it is the start of an experience like no other.

But man, I am nervous. Sure, I've got this nifty all-access press credential dangling from my neck–at least to the degree MSG grants anyone an "all-access" anything–but freely entering the Knickerbocker locker room? Asking Carmelo Anthony why he never plays to his potential? Retracing Ewing's fade-away steps on the same sacred hardwood where countless painful memories were seared into the undercarriage of my subconscious?

This can't really be happening, can it?

Writing has always come fairly easily to me, but sarcasm-laced blog posts, though mildly entertaining to me, are not exactly preparatory for navigating a team beat. Whether due to certain tabloid writers' infamous paranoia, the obsessive shadowing (both virtually and physically) by the team's p.r. staff or that general lack of knowledge of my ass from my elbow, this must be what they mean by "trial by fire."

Thankfully, I don't seem to be the only new guy 'round these parts. Just blend in, avoid writing anything too controversial right off the bat, and maybe, just maybe, people will think you know what you're doing. Plus, this is a start of a new season–one cut short by a completely unnecessary labor stoppage, but still–and it is a time for optimism, fresh perspectives and (gasp) … hope!

That's the funny thing about hope, though; it leads to expectations. And expectations, as we all know, are hardly ever met.

On this day, however, for a little while, anyway, New Yorkers are able to shed those well-worn cloaks of cynicism. Holiday-fueled optimism is (mostly) on full display at the Garden during the season-opening tilt against the Celtics but wrapped up in the shiny Christmas victory package may be something altogether more wonderful–legitimacy.

You can see it in how the Knicks are actually giving it their all on defensive rotations. You can sense it every time Carmelo Anthony lines up another "no, no, no, no, no … yessssssss" three-point attempt. You can literally feel it in your Elven slippers as the Garden shakes under the weight of the fans' raucous Christmas cheers. You can even hear it in perpetually downtrodden Mike D'Antoni's voice during the post-game presser.

This looks like a legitimate basketball team residing in the World's Most Famous Arena. And soon enough, everybody else will see it, too.

Right?

March 11, 2012: PHL @ NYK

Having all of three months of journalism experience under my belt leads me to conclude that watching, analyzing and writing about basketball may not be enough for Knicks fans. Sure, reporters are supposed to be impartial observers–bringing unvarnished information to their readers–but thankfully, given my utter lack of training, I am under no obligation to play by the rules. (And yes, calling myself a "reporter" may be taking liberty in and of itself, but a newspaper pays me to write, so screw you.)

It seems obvious that the business of professional sports demands that players, coaches and owners almost never say what they mean. This leaves fans in quite the quandary, so who better than me to act as de facto translator? The following translation comes from the March 11 post-game press conference.

Coach Mike D'Antoni: We hit a snag and we seem to whither and for whatever reason we haven't overcome any kind of obstacles. It's probably spirit more. It just seems like a two-point deficit and the world is caving in. We've got to play harder, play with much more urgency. It should always be a team game; I think Denver made that case, a lot of people's making the case. [On sitting Amar'e and Melo during the fourth quarter yesterday], I just didn't feel like it was fair to them to take them out and try something else. If guys are going to fight that hard and bring you back, I [want to] give them the chance to finish it up.

TRANSLATION: *When the going gets tough, we get going. And by going, I mean disappearing. Seriously, though, I mean, Jesus, we go down a freakin' bucket, and all of a sudden, the ship be sinkin'. We have to play harder, more urgently. You know, for once. Look, when we traded the entire world for He-Who-Must-Not-Be-Named [cough... Melo], the team that traded He-Who-Must-Not-Be-Named [cough... Melo] got better. They're still better. George Karl's... wait, I mean the former coach of He-Who-Must-Not-Be-Named's [cough... Melo] throat practically fell out and the guy's still got a better record than I do. Why in god's name would I play my "superstars" when my 2nd round draft picks play with more heart and possess more basketball IQ?! I may be dumb, but I ain't stupid.*

Carmelo Anthony: When we get down, it just seems like our whole energy just goes out the door. Times where we went on a little spurt, where we didn't make shots, guys were putting their heads down, and they heard a couple of boos out there, that just sucked the whole energy out of everybody tonight out there on the basketball court and we can't have that. Our whole energy is low. We've got to find a way to get it done right now. It's coming down toward the end of the season. Our backs are against the wall and we've got to figure it out. It's not fun though. [On sitting during the fourth quarter yesterday], I don't really know what was Coach's mindset. Maybe he was trying to save us for tomorrow. I'm not sure. You're going to have to ask him. We've just got to figure it out. Now is not the time to drop our heads and start thinking too much. We've got to figure it out, one way or another. Something's got to happen.

TRANSLATION: *When I am not getting my shots, in the places I like to get them, when I like to get them, as often as I like them, my energy just goes out the door. How dare my teammates put their heads down when shots are not falling?! That's my fucking job! And these fans, when there are boos out there, it just sucks the energy out of everybody and we can't have that. Did people not see my dominant first quarter? I'm out there on pace for 60 points and it's still not good enough it seems. The clock be tickin' on this season. This was supposed to be fun. I was told this would be fun. Where is the fun?! Y'all gotta talk to D'Antoni about the minutes. Speaking of a shortage of minutes, did MDA's contract expire yet? Now is not the time to drop our heads and start thinking too much. You only do that when you absolutely cannot stand your situation and you're angling to get out. I mean, so I've heard. From a friend.*

Amar'e Stoudemire: Yeah, it's definitely not easy. I think we could have put up a better fight [today]. We have to do a better job at knowing the importance of games. This is [losing] we've got to not get used to. I came here for a reason; I came here to win. That's something that's still close to my heart. So we've got to find a way to re-establish ourselves and implement our will on opposing teams. Tough game [tomorrow] in Chicago. A lot of games within this month against tough, competitive teams. So it's up to us as players to want to do better out there.

TRANSLATION: *Yeah, it's definitely not easy getting along with other dudes making tens of millions of dollars to play basketball. I think we could have put up a better fight [today and every other day we take the court]. We have to do a better job at, you know, like doing our jobs and stuff. I came here for a reason. And by reason, I mean $100 million dollars. And also winning. That's something that's still close to my heart. Like fashion. Fashion is close to* *my heart, too. That's just how I roll. We've got to find a way to stop letting opposing teams impose their will on us. 99 percent chance we lose in Chicago, amirite? Lots of games coming against teams that are better than us, although that's pretty much every game, but still. It's up to us players to stop worrying about getting our coach fired, start practicing free throws and stop tweeting pictures of large derrieres.*

J.R. Smith: [On being fined $25k by the league for inappropriate Twitter picture.] I didn't know it was going to be as big a deal as it is now. I definitely regret it. Anytime I do anything that takes away from our team, I'm definitely going to regret it. I can't worry about off-the-court stuff; our issue right now is on the basketball court.

TRANSLATION: *I didn't know it was going to be as big. I mean, the full body shots she sent me beforehand must have been photoshopped. I definitely regret it. Anytime I do anything that takes away from my absolutely stellar reputation—and street-cred with the ladies—I'm definitely going to regret it. For at least 30 seconds. I can't worry about the off-the-court stuff. Heck, I am not sure that I can worry about what's happening on the court, either. That stuff is way above my pay grade, ya'feel me?*

Jeremy Lin: In the third quarter, I guess, you know, they hit a lot of shots. We were a step slow on defense and I didn't take care of the ball, so they get easy run-outs in transition and then we didn't box out the way we should have, so more easy baskets and they get into a flow, they get into a rhythm. I'm getting used to being the primary or the top one, two, three things on the scouting report and I'm seeing a lot of different people try to take away what I was comfortable with. I need to make sure I stay aggressive and I don't change anything in terms of my approach. I can make better decisions, but I can't change my approach.

TRANSLATION: *In the third quarter, we imploded. You know, like we do every single game. We were a step slow on defense; stop me if you've heard this one before, and I didn't take care of the ball, so they absolutely took advantage of me and then we didn't box out the way we should have—again, stop me if you've heard this one before... [Deep breath] –so they got into a flow, they got into a rhythm. Normally I love flow and rhythm. Since Linsanity began, Landry has been taking me to all the clubs. I'm getting to be a pretty good dancer. Like Victor Cruz always says, "If you haven't got flow and rhythm, you haven't got anything." I'm considering doing the next* Dancing With the Stars, *but ONLY if they let me dance with Victor. Opposing teams are really picking on me now, though. It's shocking, really. I mean, I'm a really nice guy, what did I ever do to them?! The nerve of them taking away my right hand. My right is my bread and butter out there! I can make better decisions, but I can't change my approach. And yes, I know that my approach leads to me make bad decisions, but until someone can explain that whole chicken or egg thing to me, I'm gonna stick with what got me here.*

March 26, 2012: MIL @ NYK (one hour before tipoff)

The Empire State Building. The Statue of Liberty. The Brooklyn Bridge. Each an iconic

New York institution with a style and grace all its own.

And then there is Walt "Clyde" Frazier, whose name remains as synonymous with Gotham as it was nearly 40 years ago, when the Knicks last raised a championship banner to Madison Square Garden's rafters. Frazier, the point guard on the only two Knicks teams to ever win an NBA title, he with the career per-game-averages of 18.9 points, 5.9 assists, 6.1 rebounds and 1.9 steals, is sitting right in front of me. I try to remain calm, but sure he knows a neophyte when he sees one. Of course he does. For how else to explain how effortlessly he is able to diffuse my tension?

The first thing you notice, beyond the fact that Frazier inexplicably manages to pull off a sport coat presumably made of Del Boca Vista clubhouse drapery, is how he exudes confidence. Everyone knows Clyde's fashion lights up the room, but it is more his attentiveness and sincerity that are so immediately striking.

One gets the distinct impression that Frazier still loves the game and enjoys staying connected to it, but he seems saddened by the way the players comport themselves nowadays. And that makes me sad.

"*These guys aren't students of the game,*" he tells me. "*They may know me (and the other greats) vaguely, but I can name all of the players that came before me. The Knicks players now don't even know that their franchise was dominant in the 70s.*" I am not altogether surprised to hear Frazier point this out, but part of me now worries that he will say something during the interview that I will have no choice but to publish. Something controversial. Something unplanned. Something that won't fly in House Dolan.

Perhaps that is why Clyde never got into coaching. As is the case for so many elite athletes, the ability to instill the very same qualities that made them great in another player can be elusive. Not that the Knicks' organization ever asked him to try, of course. "*I would have only wanted to coach in New York, and I never got the opportunity.*" All the blame doesn't rest with management, either. Not one player on the Knicks' roster has reached out to Frazier for guidance. Not Jeremy Lin, (whom Frazier thinks is for real) not rookie Iman Shumpert, not struggling Toney Douglas, and not the reportedly erudite Baron Davis. Anyone else might feel slighted, but Clyde, who spends his off-seasons working on his St. Croix estate (replete with rental properties) and sailing around the Caribbean, seems to take things in stride.

Weaving my way into and out of various subjects, recently deposed coach Mike D'Antoni (who ironically ended up needing a week or two in St. Croix after his departure) finally came up. "*I was surprised,*" says Clyde. "*I liked him and was sad it didn't work out for him.*" That is not to say Frazier was sold on D'Antoni's approach. "*My rings are predicated on (defense), and I am indelibly etched in that style. Offense is capricious. Some nights the ball just won't go in, but when you bring it on the defensive end, when you're not shooting well, it can carry you. Every championship team needs a blend [of offense and defense].*"

On the oft-criticized Carmelo Anthony: "*He's one of the most prolific scorers I have ever seen. Sometimes his shot selection is poor, but I'm as baffled as every one else as to why he's struggled this season.*" Clyde didn't read too much into Anthony's controversial recent comments about his effort, either. "*Look, he makes $18M a year, but when you're on the court, it's about pride. It has nothing to do with problems with the coach; it's about what's inside you during the heat of battle. I have never felt like Melo was dogging it or that he wasn't giving it 100 percent.*"

As for the Knicks' other superstar player, the injured Amar'e Stoudemire, Frazier seems to have developed a fondness for the man who brought legitimacy back to the franchise by signing with the Knicks in 2010. "*It's devastating,*" says Clyde, lamenting the loss of Stoudemire, "*I get sad about it. I can't believe this has happened.*" Ironically, Frazier once suffered from a bad back during the 1975-76 season, missing over 30 games, but he credits yoga for a speedy recovery and says he hasn't had a problem since.

Despite widespread sentiment, Clyde actually believes that Stoudemire and Anthony can be successful together. "*[If] they want to. Like with me and Earl the Pearl, they said it would never work, but we made it work.*" Prescient advice? Conventional wisdom suggests that Stoudemire and Anthony are too similar, that they occupy the same floor space, that they are more stubborn than complementary. Frazier doesn't buy it. "*They have to check their egos to ensure that the team can succeed. That means playing in the team concept, sacrificing and playing defense.*"

Maybe there is a reason, after all, why neither of the Knicks' max-salaried players has sought Frazier out for advice. They might not like what they hear.

* * *

It is difficult to put into words what covering a professional sports team is like. On the one hand, having access to players, coaches and executives is thrilling and informative. On the other, straddling the veil between fandom and journalism–out of sheer necessity, mostly–eradicates any notions of childhood nostalgia still lingering deep within.

Let's be honest; there are few good guys here and even fewer "heroes."

In New York, Knick loyalists must endure a tyrannical owner, hell-bent on destroying mankind, or, depending on who you talk to, merely his fan base. The team's best player is a flawed enigma, maddening to watch, possibly beyond reclamation, and apparently not the least bit preoccupied with shedding negative perceptions of him. Costs to attend games are out of control, to the point that average families–especially those with children, those same children who would presumably support the franchise going forward–have been priced out.

And despite the gut-wrenching losses, the decades of championship-drought, the mismanagement, the haphazard manner in which this franchise comports itself, I still consider it an honor and a privilege to bring my wacky opinions and analysis to Knicks fans. The names and faces may change–not Dolan's, obviously, he is our one true constant; our rock–and the losses may continue to pile up, but for those who bleed orange and blue, we wouldn't want things any other way. Well, actually, we would, but still.

THIRD QUARTER:
FEBRUARY 29—MARCH 15

Of course it was too good to last.

Had we learned nothing? Of course this miraculous streak of beautiful D'Antoniball was destined to go up in flames.

And the cauldron in which all our hopes and dreams burned was such a typically New York Knick-y kettle, it was as if it was pre-ordained by a divine being with a truly sick, sadistic sense of humor.

All of the collective fears/worries/self-fulfilling prophecies about how Melo would adjust to this New Knicks World Order came horribly true. Whether it was due to struggles with the assortment of injuries that had strapped him to the bench during the height of Linsanity or because he was just pouting and breaking offensive sets at the indignity of having to play second fiddle to a new, far more adored star, we'll probably never know. What we do know is that Anthony's shot was as broke as a joke, and that the fans were beginning to turn on him. Compounding the Knicks' PR woes, Amar'e Stoudemire—still working to get his back right and his game straight—was struck by a tragedy of unimaginable proportions when his older brother, Hazell, was killed in a car accident in STAT's native Florida. Meanwhile, Lin's play was beginning to quake slightly from its Herculean apex. The defense, which had actually carried the team during their brief era of good feeling, came staggering to a halt when Melo and STAT were squeaking parquet together.

And thus what had been for the briefest of moments a beautiful, coherent team, had seemingly reverted to the cluttered, ill-fitting mess of the first part of the season, in which every man on the roster looked like he'd far prefer to undergo extensive reconstructive dental surgery than spend another millisecond with his teammates on court plying his trade.

Surveying this bleak, ruined, dystopian landscape, Mike D'Antoni abandoned ship, quitting as head coach of the Knicks to in all likelihood drown his sorrows in whatever café in Perugia would have him and eradicate every brain cell that contained a memory of his tenure in the Big Apple. For a fan base that had whipsawed into a state of near-frenzy, yet another descent into that all-too-familiar state of despair was somehow an entirely expected turn of events. Expected, yet equally heartbreaking. Once again, all seemed lost (again).

For the moment...

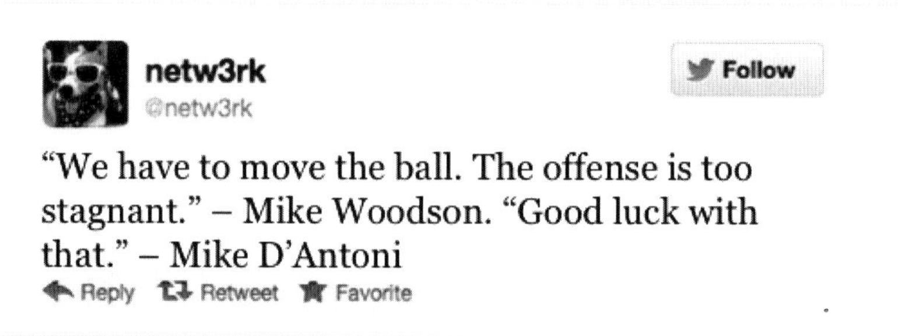

Amar'e's take on the era: "Phenomenal."

A psychiatrist's diagnosis of Knicks fans: "Querulous Paranoia"

#1 single in Finland: Loreen, "Euphoria"

Toney Douglas's Google search history: "cool neck towel fashions"

From the papers:"I think anytime you go from the early part of the season, just having the ball and me just having the ball and being the distributor, and now just running the wings and waiting for the ball to come to me, that's quite an adjustment for myself." (Carmelo Anthony, *The New York Times*, 3/6/12)

Key stats: Mike Woodson went 18-6 in 2012. If he could keep up that pace during a full regular season in 2013, the Knicks will win 62 games.[47] Dude!

47 Hopeless optimism Copyright 2012 by Mike Kurylo.

CHAPTER 8
AN OPEN LETTER TO JAMES DOLAN

BY ROBERT SILVERMAN
KnickerBlogger.net

The Editors would like to state in no uncertain terms that they do not in any way endorse the following claims made by Mr. Silverman. In fact, we feel that he is less qualified to coach the Knicks than every other multi-celled organism on the planet. Alas, after decades of Knick fandom, the soul-crushing defeats and the inexplicable stupidity exhibited by the team's so-called brain trust, Mr. Silverman's mental health has clearly been compromised. We do feel though that it is important to include the witless blatherings of an addled mind like his if only to provide proof of the derangement that afflicted the fan base in varying degrees in the midst of one of the most maddening, surrealistic Knick seasons in recent memory.

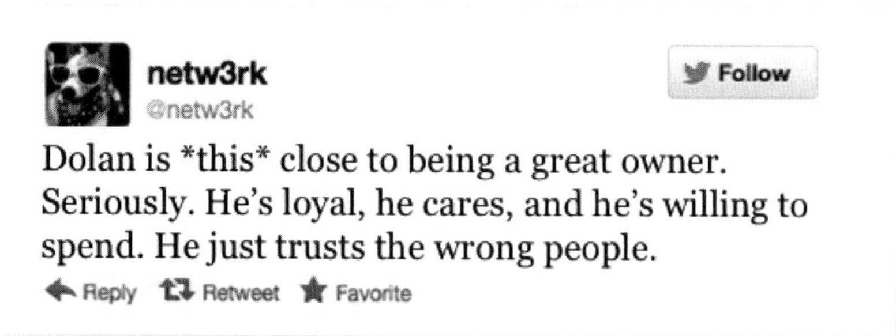
netw3rk
@netw3rk
Follow

Dolan is *this* close to being a great owner. Seriously. He's loyal, he cares, and he's willing to spend. He just trusts the wrong people.

Reply Retweet Favorite

Dear Mr. Dolan,

Can I call you James? Great. Listen, Jim, if you aren't too busy lavishing premature laurels upon your first place hockey team, you may have noticed that the other denizens of your fine bit of real estate on top of Penn Station aren't doing very well these days. And as is oft the case, New York fans want the head coach of said offending team served to them on a silver platter with a nice sauce béarnaise.

I'm here to tell you that the madding crowd is in this instance one hundred and twelve percent right. Mike D'Antoni must go! Fire the offending cretin immediately! Send him back to Arizona where he'll be forced to show his ID[48] every time he's caught lurking in the bushes outside Steve Nash's house. And rightly so. That 'stache he sports

48 Arizona SB-1070. Research it and make your own decision.

is very *foreign*-looking, if you catch my drift. Who knows what all those years toiling in exile with those blasted European *socialists* has wrought upon his formerly finely tuned offensive mind. Now that the scurvy dog has been shown the door, you'll need a captain for this leaky ship. Lucky for you, I've got the perfect candidate.

Me.

Granted, I have never coached or played organized basketball at any level. Truth be told, I couldn't put the ball through the hoop if you gave me a three-foot stepladder. But that's precisely the point! Just as is the case in the current political climate, where the hoi polloi is clamoring for an "outsider" who hasn't been tarred by the all-corrupting influence of the D.C. establishment, Knick fans will rally behind me, a man who is completely unqualified for the job at hand. C'mon, think outside the box! After all, isn't "conventional wisdom" the oxymoron that led to this muddled mess of a roster in the first place? Why not hire a visionary who, days after the trade for Carmelo Anthony went down, wrote:

"I know that in a couple of years we're going to regret this ghastly trade. Just like the Francis deal or the Curry deal or the Marbury deal and on and on, this is going to a be a franchise-altering mistake."

You may be thinking, "But Bob, how will your sage-like wisdom improve the Knicks right now?" Easy peasy, Jimmy. If you glance at my resume you'll note that I have spent many a year writing, directing and acting in stage plays. If anyone is qualified to manage a group of whiny, pouting, narcissistic, me-first divas (differing sexual peccadilloes notwithstanding), it's yours truly. Plus, the same basic ingredients that make for great theater make for great basketball–rhythm, tempo, floor spacing, unselfishness, and possibly having someone in a tiger costume show up in the second act.

More importantly, let's get real, Jimbo. Unless Cablevision scientists have figured out a way to bring Red Holzman[49] back from the dead (perhaps the necromancer who revived Mike Bibby can look into that), there isn't a coach alive that would be able to pull this so-called team out of the gutter. Instead, we must change the narrative! In America, we don't sell the steak. We sell the sizzle! In place of the justly deserved groans about Amar'e and Melo being unable to coexist, pinning all our hopes on the overburdened back of Baron Davis, or insisting that Jeremy Lin can return to his state of grace, get Democrats and Republicans to work together, and get the kinks out of that whole "Unified Field Theory" that they've been futzing with for decades, let's make the story about me, the deranged lunatic who will enter a pre-game press conference gnawing on an alley cat found outside the Sbarro's next to the Garden and bellowing, "To defeat those rancid cowards from Charlotte I must consume the beast's life force! Bring me the still-beating, lipid-clogged heart of Boris Diaw!!"

49 Holzman coached the New York Knicks from 1967-1982, won two titles, and entered the Hall of Fame in 1985.

Lest you think this is all a tired rehashing of the plot of the movie *Eddie,*[50] unlike that piece of saccharine Hollywood chum, I'll bring the weird, Jimm-o. I guarantee that I'll do everything in my power to distract the yokels from the execrable product you're putting on the floor. Think of the possibilities! Those ink-stained wretches will barely be able to contain their glee as they scribble bile-filled jeremiads about your incompetent ignoramus of a coach who refuses to stop racking up technical fouls for chain-smoking on the bench during the games, publicly refers to Iman Shumpert as ephebe, quotes liberally from Saul Alinksy's *Rules for Radicals* and as far as my plans for the Knick City Dancers go, well… let's just say it's not fit to print in a family newspaper. We may not win, but at least we won't be boring. Isn't that preferable to the pall of all-encompassing, inevitable doom that's befallen the Knicks, where fan and reporter alike hover like jackals, ready to pick at the rotting bones of this mangled corpse of a season? In these dire end-times, you need as wily a political operative as the great Rowdy Roddy Piper, who once said, "Just when they think they've got the answers, I change the questions."

I'm ready to go to work, boss. Get the *Times* on the blower when you're ready to embark on this magical journey. They'll know how to get in touch with me. Better yet, stand in the middle of a grassy field (any field will do), flap your arms wildly, let out a joyous, Whitman-esque yawp and I'll come-a-runnin'.

Signed,

Your Future Head Coach,

Robert Silverman

---

50 A feature film starring Whoopi Goldberg and Frank Langella and a slew of NBA luminaries in which a fan (Ms. Goldberg) is hired to coach the Knicks and improbably leads them to the playoffs.

DON'T FORGET ABOUT STEVE NOVAK AND J.R. SMITH
(A.K.A. THE ODD COUPLE)!

BY SETH ROSENTHAL

It wasn't easy to separate J.R. Smith from a basketball. It was well known that J.R. loved to dribble in circles and find his own shots, often to a maddening extent. Smith wasn't a bad passer, but he was wont to decline opportunities to share out of reluctance or just sheer neglect. J.R. seemed to fancy himself a surefire scoring option, so it was difficult to imagine him going out of his way to feed a teammate. It happened, though. It happened a lot, but only for one man.

Steve Novak's 47 percent three-point shooting for New York did not go unnoticed by our friend J.R. Novak's mere presence on the floor transformed Smith's hoggish ways into a bizarre and occasionally detrimental streak of hyper-deference. Knicks fans were bewildered to see Smith eschew clean shots and curl away from open lanes to find Novak as he squared up behind the arc. J.R. was so committed to the set-up that he'd test the laws of physics, attempting to navigate imaginary angles or pass through solid human opponents. Often, he'd toss the skip pass, and then raise his arms aloft in celebration before Novak even attempted his shot. Some of J.R.'s turnovers could probably be attributed to spotting a tall pale spectator in his periphery and acting on reflex.

J.R.'s uncharacteristic deference to Steve is difficult to quantify, but here's a hint: With Novak on the bench, Smith averaged 2.3 assists per 36 minutes. With Novak on the court, that number jumped up to 3.6 per. Looking at it the other way, Novak hit 49 percent of his threes with Smith on the floor and just 40 percent without him.

Perhaps the most wonderful example of the Smith-Novak dynamic came on April 17, when Smith drained seven three-pointers—two off the Knick single-game record—in the first half alone. Against all odds, Smith hardly even attempted to break that record in the second half, instead stepping aside as Novak surpassed him and finished with eight three-pointers to his seven.

It was a sight to behold—a symbiotic relationship based on reverence, it seemed. One gunner bowed to another, and it paid off. The Knicks employed many fine tandems, but none as lovely as this one.

CHAPTER 9
MIKE D'ANTONI: AN OFFENSIVE RETROSPECTIVE

BY MIKE KURYLO
KnickerBlogger.net

So Linsanity came and went, a phenomenon that transcended sports, which on its own would have made the 2012 season notable. But the Knicks weren't nearly done in making their mark in history. Next up would be swapping coaches. Changing the man in the suit on the sideline may seem like an uncommon occurrence, however, it's the Gregg Popovichs of the world that are extremely rare. In Darth Gregg's 16 and a half-year tenure in San Antonio, the Knicks have had 8 different head coaches. I thought that D'Antoni deserved a send-up for his three and a half seasons on 32nd street. Unfortunately, what best defined D'Antoni was more who he was prior to New York.

From the moment the Knicks hired Mike Woodson as their new assistant coach, it didn't take a genius to figure out that Mike D'Antoni's days in New York were numbered. Forgetting for a moment Woodson's Hoosier ties—both Isiah Thomas and General Manager Glen Grunwald attended Indiana University—on the surface at least, the two coaches appeared polar opposites: the head coach sporting neatly combed silver hair and impossibly bristly lip coif, while his trusty assistant donned a hairless head so shiny you could ice skate on it.[51] D'Antoni lived and breathed offense; Woodson's sole contribution to the craft was throwing the ball to Joe Johnson and hoping for the best (a 21-foot jump shot, usually)–while ostensibly preaching defense.

On Wednesday March 17, just about the midway point in the season, Mike D'Antoni

51 We are in no way condoning skating on Mike Woodson's head.

NEWYORK
16

resigned as coach of the Knicks. With Woodson at the helm, the Knicks won eight of their next nine games–and in convincing fashion, averaging a 14.3-point margin over that span. Under Woodson, New York embraced a defense-first philosophy, with their subsequent 18-6 record serving as sweet vindication for those who longed for the days of Ewing, Oakley, Starks, and Mason. Where Mr. Pringles had failed, Mr. Potato Head was surviving—and thriving.

But enough about spuds. To tell the full story of the rise and fall of Mike D'Antoni, a little bit of history is in order. I've always been fascinated by offensively skewed teams. The late 80s and early 90s were such a great time for me as a sports fan, given the offensive explosions in both baseball and football. Mark McGwire and Jose Canseco were making the long ball sexy in Oakland, ushering in baseball's second power-ball era. In 1991, Cecil Fielder became the first player in 14 years to reach 50 home runs, and his Detroit Tigers finished second in the AL East with a starting rotation led by Bill Gullickson and Walt Terrell–two guys that would have trouble getting outs against a wiffle ball team.

Around the same time, the run and shoot was making its frantic presence felt in the NFL, most notably with the high-octane attacks of the Houston Oilers and Detroit Lions. The offense may have been controversial at the time, as power football was still very much the system of the day. Removing the tight end and fullback from the field (and often from the roster) and using four receivers to spread the field made for a brand of football that was nothing if not a blast to watch. I was so excited by the run and shoot that I purchased season tickets to the talent-challenged World League of American Football's New York-New Jersey Knights, whose head coach, Mouse Davis, was an early architect of the run and shoot. The World League would fold after two seasons.[52]

But the entertainment factor wasn't the only reason I was drawn to these styles of play. Teams that are stronger on offense typically get an unfair amount of criticism thrown their way. Clichés like "great pitching beats great hitting" and "defense wins championships" are treated as fact, though they remain largely unsubstantiated. Great offensive teams won championships too, a fact that would seem obvious to any Yankees fan. After all, how far down the list of great Yankee players must one go to find a world-class pitcher? Even the team's nickname, the Bronx Bombers, implies *offensive juggernaut*.

To the masses, defense is seen as yeoman's work—something that requires effort, and not the kind that inspires much in the way of Homeric odes. That's why players that can score but have trouble defending (somewhere, Carmelo Anthony chugs a gallon of chocolate milk[53] at halftime) are often viewed as lazy. But this kind of thinking ignores the fact that a

52 Technically it would revive itself as the World League, NFL Europe and NFL Europa before ultimately folding in 2007. Nevertheless the games were really bad to watch, yet I still watched them.

53 Two studies at the University of Texas at Austin in 20111 showed that chocolate milk outperformed sports drinks in terms of recovery and building muscle.

good offensive player has to put in countless hours of work (thousands and thousands and thousands) honing his or her craft. Fans define Ben Wallace as "hard working" because of his spectacular ability to rebound and defend, while completely ignoring the fact that he's probably spent as much time working on his jump shot or his low post game as I have sampling mall counter makeup. And it's precisely this distorted definition of defense, reinforced by high-octane collapses and notions of middle class work ethics, that ultimately allows for such labels. It's easy to appreciate the philosophy behind preventing scores being more important than creating them. But that doesn't make it an accurate reflection of reality.

Which brings us back to D'Antoni. Given my scoring fetish (take that how you will) I was ecstatic when the Knicks hired the former Suns skipper in 2008. With D'Antoni as coach, Phoenix had finished either first or second in the NBA in offensive efficiency from 2005 through 2008. When the Knicks hired him, he had a winning percentage of .608.[54] Clearly his offense wasn't just a gimmick; it actually worked and worked really, really well.

Much like the NFL's run and shoot, D'Antoni's offense, a.k.a. *Seven Seconds or Less*, as legendary NBA scribe Jack McCallum termed it, was highly criticized. Despite averaging 58 wins over a four-year span, D'Antoni's Suns never advanced past the Western Conference Finals. As such, it was obvious to his detractors—and to far too many pundits and commentators—that his teams simply didn't play defense and thus were doomed to a kind of title-less purgatory.

But it's not as if D'Antoni's Suns were a defensive void. Despite finishing close to or last in points per game allowed during each of D'Antoni's four full seasons, Phoenix was actually an average defensive team efficiency-wise, finishing between 13th and 17th, when adjusted for pace. The Suns made the conference finals in consecutive seasons and likely would have done it three times in a row. Unfortunately, the 2007 playoffs were lost for Phoenix when Robert Horry's Byfuglien-esque[55] hip-check on Steve Nash resulted in suspensions for Amar'e Stoudemire and Boris Diaw, who both left the bench. Even with a depleted roster that would have had most coaches throwing up the white towel (losing Joe Johnson to free agency and Amar'e Stoudemire to a devastating knee injury being the two most obvious examples), D'Antoni's teams remained a force to be reckoned with in the Western Conference.

54 D'Antoni's record was 267-172.

55 Somewhere there is a dictionary where the definition of "hockey goon" simply has a picture of Dustin Ray Byfuglien. A Canadian dictionary.

That the Suns failed to win a title wasn't proof of anything, for the simple fact that most teams *don't win a championship.* As in any sport, winning in the playoffs demands equal parts skill and luck. The Suns, by virtue of their intricate, almost jazz-like offense, had the first element in spades but were a bunch of Lucy-duped Charlie Browns when it came to the latter. Really, it shouldn't be hard to imagine an alternate Earth where, blessed with talent that was the perfect embodiment of his high-flying offense, D'Antoni ends up snagging one or two rings.

For Donnie Walsh and the Knicks, hiring D'Antoni was a sound move, since he was quite simply the best coach available. Other candidates, like Mark Jackson and Tom Thibodeau, were unproven as head coaches. Truth is, a bad coach can ruin a team much more easily than a good coach can turn a mediocre group into a winner. When choosing someone to man the sidelines, it's better to play safe than risk being sorry.

To outsiders, the '08 Knicks and the '04 Suns might've seemed similar. After all, both had been losing teams in recent years. Boasting an offense predicated more on the system than the actual personnel, it seemed almost inevitable that D'Antoni's success would continue. But unlike with Phoenix, New York didn't have a slew of future All-Star forwards on the roster. Nor did they boast the league's best pure point guard, as the Suns had with Steve Nash. Instead of Nash, Marion and Stoudemire, D'Antoni had Chris Duhon, Wilson Chandler and David Lee.

For two seasons with D'Antoni patrolling the Garden's *Titanic*-like sidelines, Donnie Walsh could do little more than rearrange the deck chairs. Truth be told, it would have been difficult for anyone to find complementary pieces for a roster core that might've struggled to win a D-League title. Wilson Chandler, Chris Duhon and Al Harrington would've been complementary pieces for even the most middling of NBA teams. For the Knicks, they were cornerstones. That was fucking terrifying.

Roster blight be damned, D'Antoni brought his fast paced offense to a city well versed in hurry. But without the kind of quality players that defined his tenure in Phoenix, D'Antoni's new charges struggled, winning just 32 games the first year, and 29 in the 2009-10 campaign. New York's offense, touted as Suns redux by many in the media, was pretty mediocre ranking 17th in offensive efficiency in each D'Antoni's first two seasons.

Recognizing his team's glaring lack of punch, D'Antoni tweaked the offense to run through the Knicks' best player: David Lee. For D'Antoni, who had built his reputation on a point-guard-heavy attack, this was indeed quite the departure. Still, he was flexible enough to realize that he lacked the necessary Nash-ian floor general to make everything tick. Duhon was the team's highest minute getter in 2009, but by late 2010 it was apparent he wasn't *the* answer, let alone *an* answer. As the 2010 season ground on, Duhon would rack up quite the collection of DNP-CDs,[56] as D'Antoni tried a host of replacements. However Toney Douglas, Tracy McGrady or Sergio Rodriguez did not manage to wrest the job away.

By the summer of 2010, with the losses quickly mounting, a growing contingent of D'Antoni's critics lamented the team passing on Mark Jackson as head coach. Jackson, conventional wisdom had it, would've at least been committed to that halcyon pillar of New York basketball: defense. Mark Jackson, it must be noted, whose defense as a player could be best described as matador-like. Jackson stuck around the league for 17 mostly productive seasons because of his passing, so why would anyone believe that he would become a defensive expert as coach? [Note: when Jackson was actually given a head coaching position, his 2012 Golden State Warriors finished 27th in the league in defensive efficiency.]

That same summer would give the NBA the villainous troika known as Miami Thrice (LeBron James, Dwyane Wade and the ostrich-human hybrid known as Chris Bosh). The Knicks, who many thought had the best chance at landing the Swiss Army knife James, managed to reel in Amar'e Stoudemire, Raymond Felton and Angus T. Jones[57] look-a-like Timofey Mozgov. With rising prospect Danilo Gallinari and second round rookie surprise Landry Fields, the Knicks appeared to be headed towards respectability for the first time since the Clinton administration.

After a win over Atlanta on February 16, 2011, New York stood two games over .500, which was a marked improvement by just about every metric. However, no amount of potential or promise would prove enough to stave off the inevitable, as days later the team traded Gallinari, Felton, Chandler and Mozgov for Carmelo Anthony, Chauncey Billups and Anthony Carter. Despite the midseason roster upheaval, the Knicks made the playoffs for the first time in what felt like a geological epoch.

D'Antoni's third season had been a resounding success for a franchise grown accustomed to their summer vacations starting a few weeks after Easter. The offense, perhaps the best measure of D'Antoni's ability as coach, ranked seventh in the league. Even after

---

56 DNP-CD stands for, "a player that was too shitty for the coach to put on the court," or, "Did not play, coach's decision." Both are applicable.

57 Angus T. Jones is an actor in the show *Two and a Half Men*. He's the "and a half."

a first round sweep at the hands of old nemesis Boston, the team had high hopes for the future. Six months later, the lockout's worst-case scenario narrowly avoided, the Knicks managed to pull off an 11th hour coup by inking veteran defensive anchor Tyson Chandler to a max contract. Finally, after years of roster shape shifting, the Knicks were good enough to compete in the Eastern Conference. Leading the charge was D'Antoni, who was now expected to make the most of his opportunity.

Fast forward to mid-February. D'Antoni's Knicks have just outgunned the Cavs 120-103, bringing their record to 18-18. It's a respectable mark, to be sure. But then the floor dropped out completely, as the Knicks lost six straight over the next few weeks–a spell that would end with D'Antoni being formally, finally ushered away from the Garden sidelines.

For someone who'd spent a significant chunk of his life as a fan championing the cause of *offense uber alles*, I can probably conjure a couple of excuses for D'Antoni's failure. Take, for instance, the time—or the timing—of his resignation. Every NBA team has parts of its schedule that are much more difficult than others, and D'Antoni's final six games, in which the Knicks played five on the road and five against playoff teams, were no exception. Firing a coach for one bad streak might seem, for most fan bases, something of a knee-jerk reaction.

More importantly, D'Antoni never had a stable roster with which to work; his first two years were rebuilding years, a painful transition out of the failure-pile-in-a-sadness-bowl that was Isiah Thomas' tenure. His third season saw the team identity change literally overnight with the roster-gutting Carmelo Anthony trade. The 2012 season was shortened by the lockout, and thus hadn't had a full offseason to work with the team. If you think about it, during D'Antoni's New York tenure, he had a stable roster for less than a single season.

Sadly, despite my best rationalizations, and my desire for the Knicks to be the poster child of "offense wins championships," D'Antoni had failed the Knicks in multiple areas. Although the team was decent defensively, the offense was in shambles. This was a bad sign for D'Antoni, who had come to pretty much define what a 21st century NBA offense was supposed to look like.

And what about player development? D'Antoni, a point guard himself as a player, didn't recognize the talents of Jeremy Lin—that is, of course, until he was forced to do so. The future phenom would sit on the bench for weeks watching Bibby, Baron, and Douglas attempt in vain to run the point. The Knicks suffering through these point guards was like Nirvana suffering through Chad Channing[58] and his aversion to songwriting, when David Grohl[59] was readily available. The coach that needed a point guard had a prodigy right under his broom-bottomed nose every day in practice but couldn't recognize it? It seemed impossible to believe.

58 Nirvana's original drummer.

59 For you younguns, Dave Grohl was in another band before Foo Fighters. They were pretty good too.

But Lin wasn't the only failure. Second round pick Landry Fields regressed mightily from rookie revelation to sophomore slip-up. Under the league's most "shoot it if you're open, even you Jared Jeffries" coach, Fields forgot how to shoot. He went from hitting two of every five three pointers in 2011 to hitting two of every eight in 2012. His free throw shooting dropped 20 points, and at 56.2 percent approached Ben Wallace territory.

Finally, there's the case of Amar'e Stoudemire. STAT's decline, although not as precipitous as Fields', was far more damaging when considering the economic ramifications. Indeed, the Knicks had handed Stoudemire somewhere in the vicinity of $18 million in 2012, expecting the league's preeminent pick-and-roll dive man to do what he'd built his career in Phoenix doing, namely attacking the tin. In the orange and blue, Stoudemire's scoring had become pedestrian. After leading the league in true shooting percentage[60] in 2008 at 65.6 percent, by 2012 his 54.1 percent clip was league average. After five straight All Star appearances, Amar'e was not selected for the 2012 midseason classic, while Chris Bosh—whom three years ago precisely no one would've chalked above Stoudemire—enjoyed All Star honors.

For good or ill, this all went down on D'Antoni's watch—the same coach that made Amar'e Stoudemire and Shawn Marion household names in Phoenix; the coach that helped Leandro Barbosa develop into one of the league's best sixth men; the coach that dug up afterthought Boris Diaw from Atlanta and turned him into a starter; the coach that turned a pretty good Steve Nash into two-time MVP Steve Nash.

In a season not afflicted by the lockout bug, perhaps D'Antoni would have had more time to better prepare his team. Maybe another city would have been more patient and allowed him a more stable roster to work with. Perhaps a different group of superstars would have better fit his strengths. If the Knicks didn't have a head coach in waiting, possibly they could have convinced D'Antoni to stick it out for another year. Maybe Fields's rookie season was a fluke, Stoudemire's body broke down, and no one in the world knew what Jeremy Lin had to bring to the table.

But Bill Parcells's rule applies: Mike D'Antoni was only what his record says he was. His players didn't develop. His offense didn't work. His team didn't win. And when he handed the reins to someone else, the Knicks began to win. Even I, who would have loved to cover a team that could outscore its opponents at will, couldn't muster enough excuses to see D'Antoni's time in New York in a positive light. In the end, Woodsanity had ensued and Gotham had forgotten all about D'Antoni. Not because defense wins championships, but because D'Antoni failed at the job he was hired to do.

---

60 I already gave a basic description of true shooting percentage. But if you're here you want more, so the equation is pts/(2*(FGA+(.44*FTA))). League average is about 54 percent.

CHAPTER 10
DEE-FENSE (CLAP, CLAP) DEE-FENSE (CLAP, CLAP)

BY JARED DUBIN
Hardwood Paroxysm

Defense. That's what New York Knicks fans have come to expect of their best teams over the years. The 1970s. The 1990s. From Red, Clyde, Pearl and Willis to Riles, JVG, Ewing and Camby. Coming into the 2011-12 season, defense had mostly been an afterthought or a joke to the Knicks for the better part of a decade. But that was all about to change.

For Iman Shumpert, it all started with an early January game against the Washington Wizards. After a season-opening win against the Boston Celtics, a 1-2 California road trip, and back-to-back home losses to the lowly Toronto Raptors and Charlotte Bobcats, the Knicks sat at just 2-4, and were looking for answers on both sides of the ball. Once again, the Knicks had fallen behind by double digits, as Wizards point guard John Wall seemed to get anywhere and everywhere he wanted on the court. Desperate for answers, Mike D'Antoni pulled starting point guard Toney Douglas, unleashing feisty rookie Shumpert on Wall and letting Shump do what Shump does best: pressure the living hell out of any man with the ball in his hands, and make life more difficult for him than he ever could have imagined. A 36-22 second quarter deficit slowly but surely gave way to a 99-96 comeback victory highlighted by a game winning three-pointer by Carmelo Anthony.

Buoyed by that seemingly trivial win, the rest of the season would end up being the best ever on the defensive end in over a decade for the New York Knicks–a team that a few decades earlier had built its identity on being the meanest, nastiest, no-layups-iest defense in all the land.

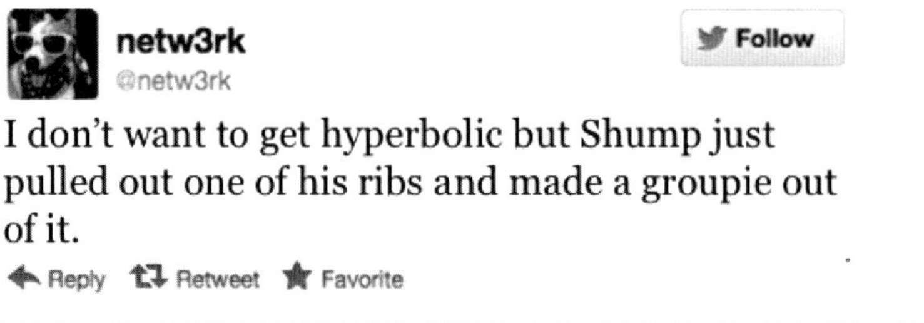

To fully appreciate just how impressive a leap New York was making on the defensive end, it's important to understand exactly where they'd been. In the previous three seasons under Mike D'Antoni, the Knicks finished 23rd (2008-09), 27th (2009-10) and 22nd (2010-11) in defensive efficiency.[61] From 2001 to 2008, they never finished higher than 15th, and that was one of only two seasons where they finished inside the top 20 (they were 18th, 23rd, 15th, 25th, 26th, 25th and 29th, in that order, from 2001-02 to 2007-08). By the end of the 2012 season, the Knicks ranked 5th in the NBA in points allowed per 100 possessions at 98.4, their best mark since the 2000-01 season (95.5), when they finished third.[62]

But let's back up a bit. Though I left him out of my cleverly crafted, in-media-res-esque lede,[63] any conversation about the Knicks' defensive improvement that doesn't begin with the wondrous Tyson Chandler isn't starting in the right place. From the Christmas Day opening tip until the last second of the Knicks' Game 5 loss to the Heat in the first round of the playoffs, the tall, strong, bearded beast was nothing short of a *monster* on the defensive end.

Most basketball players, even at the NBA level, play defense with their hands. Tyson plays it with a seamless combination of feet, brains, chest and—cliché alert—heart, and all in equal parts. At various points during the season, he swallowed up centers as disparately talented as DeSagana Diop and Dwight Howard any time they tried to come at him, be it in the post or off the dribble. For the season, Howard—the best center alive by most everyone's standards—averaged 20.6 points and 14.5 rebounds per game on 57 percent shooting. Against Tyson Chandler and the Knicks, those numbers plummeted to 9.3 points, 7.7 rebounds, and 52 percent shooting from the floor. And while Howard hoisted an average of 13.4 field goal attempts per game on the year,

61 Defensive efficiency is points allowed per 100 possessions, a better measure of defense than pure points allowed per game because it adjusts for pace.

62 Somewhere, Lavor Postell grins.

63 The answer is yes, I did so on purpose specifically so I could use that turn of phrase.

Chandler succeeded in limiting the beastly pivot to 21 shots *total* in the three Knicks-Magic games Howard appeared in.

But it wasn't just big men that Chandler thwarted. In the new defensive system designed by then-assistant coach Mike Woodson, Tyson was also called upon to pick up guards and wings off switches, often shutting down their scoring opportunities with relative ease. On 69 such possessions, Chandler held guards and wings—everyone from little Brandon Jennings to big, bad LeBron James—to just 17-of-60 from the field (a paltry 28.3 percent) and forced five turnovers while committing just *four* fouls. Normally, when a guard or wing gets a big man on a switch, they try to drive right past him and get to the rim. Not Chandler, whose quick feet, long arms and heady presence deterred even the most seasoned of lane-slashers from even thinking about that kind of gambit. Over and over again, guards met by Tyson's eerily speedy switches would settle for step-back jumpers, high-arching floaters, or winding, spinning layups, hoping in vain to loft the ball over Chandler's outstretched arms. Time and again, Chandler showed just how fruitless that idea was, leaping forward, right hand held high and straight, blocking any view of the rim. Time and again, the ball clanged harmlessly off the rim or the backboard, and into the hands of a waiting Knick.

Flip on just about any Knicks game from the 2011-12 season, and it doesn't take long to see the true portrait of Chandler's often thankless work; a crazed madman still somehow in control of everything around him. You see him barking out orders at the top of his lungs—"Switch, STAT![64]" or "You got him, Shump!" or "I got your help, Melo!"—all the while shifting, patrolling, waiting for someone, anyone to challenge his space, because if they did, they were dead. To put it bluntly, Tyson Chandler *owned* one half of the court this season. It was his and his alone, and no one could take it away from him.

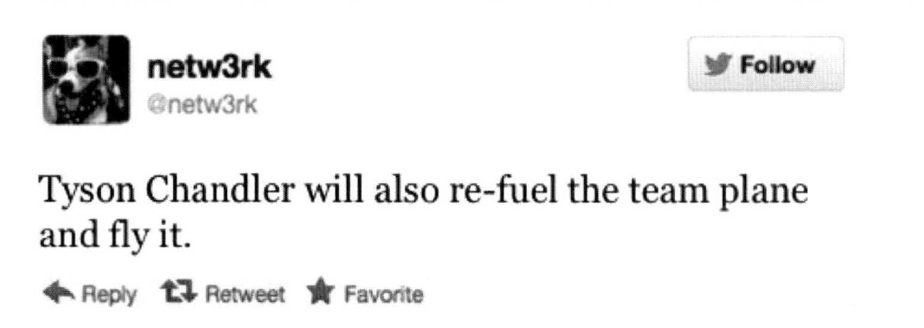

64 Amar'e's always been confused by "switch, STAT," because he thinks whoever's saying it is speaking German.

That's not to say he did *everything* himself, however. Woodson was, in the purest sense, the architect of the system that Chandler would so masterfully execute. After two years of passive resistance, Mike D'Antoni finally agreed in the summer of 2011 to bring on a "defensive coordinator." Together with General Manager Glen Grunwald, D'Antoni decided that Woodson, Grunwald's college teammate at Indiana University and the former head coach of the Atlanta Hawks, would be that guy. Almost immediately, Woodson made his presence known. Stories leaked that the abbreviated Knicks training camp was featuring an added emphasis on defense—certainly more than one had grown accustomed to during D'Antoni's first three seasons at the helm. In sessions televised on MSG, Woodson literally took over portions of practice, teaching the finer points of his defensive system. And though it took a few games for the team to get settled in, that system, and the intensity with which they practiced it, carried over into the regular season.

Through the season's first 23 games, from the Christmas Day opener against Boston to the February 3rd game against the Celtics that marked the last before that fateful home game against the Nets,[65] the Knicks allowed 98.5 points per 100 possessions, good enough for 11th in the league at the time. During that stretch, they allowed over 100 points just six times, while holding their opponents to under 90 six times.

But it was really during Linsanity that New York would discover its true defensive identity. Although those games will forever and understandably be remembered for Jeremy Lin's marvelous, out-of-nowhere ascent to stardom, really it was the New York defense that carried them to eight victories in nine contests. During that stretch, the Knicks allowed a paltry 94.8 points per 100 possessions, an improvement of 3.7 points over the previous 23 games, or roughly the difference between New York's 5th ranked defense on the season and the 15th ranked defense of the Houston Rockets. What's more, they held the opposition under 100 points in every game, and under 90 in five of them.

Unfortunately, that dreamlike stretch was followed immediately by an equally depressing 10-game stand that saw the Knicks allow 103.7 points per 100 possessions—that, had it marked their season's effort, would have ranked 21st in the league. Incidentally, the stand coincided with the return of Carmelo Anthony from a pair of nagging injuries, culminating in Mike D'Antoni's last game as Knicks skipper.

Enter "Woodsanity." In winning eight of their first nine games under the then-interim coach Mike Woodson,[66] the Knicks allowed an utterly preposterous 90 points per 100 possessions, besting Chicago's league-leading full season mark by 5.3 points. They allowed over 100 points just once, a 15-point blowout win over the Indiana Pacers that

65 Renaldo Balkman brought everyone cupcakes before the game to celebrate some dude's birthday whose name he couldn't remember. Nothing else happened.

66 Apparently he was officially given the job during the offseason. So long as Isiah Thomas is walking around, however, every Knick coach is really just an interim.

wasn't nearly as close as the final score indicated. They held seven of the nine opponents under 90 points and three of them under 80. Their average margin of victory in that nine-game span? A cushion-y 14.2 points per. They were the hottest team in the league, the one nobody wanted to see come playoff time.

And then, injuries—lots of them. Jeremy Lin was the first to go down, hurt in the 7th game of Woodson's tenure and subsequently declared out for the remainder of the season with a torn meniscus in his knee. Next came Amar'e Stoudemire, who began experiencing back spasms and was diagnosed with a herniated disk that would keep him out until the last game of the regular season. Jared Jeffries, who'd worked so hard to redeem himself in the eyes of the Knick faithful, becoming something of a cult hero, went down with multiple minor maladies. The Knicks managed to stay hot through the end of the season, but the injuries, combined with a glaring lack of depth, impacted them on the defensive end. In the final 15 games of Woodson's regular season coaching tenure, the Knicks allowed 102.0 points per 100 possessions, roughly the equivalent of Houston's 15th ranked full season defense.[67]

By the time they'd gotten to the playoffs, the Knicks—depleted and, thanks to Lin's season-ending sidelining, deflated—had no hope of containing LeBron James, Dwyane Wade, Chris Bosh and the Miami Heat. Of course, it didn't help that no one on the planet can guard James. Similarly, whatever slim chance the Knicks had of containing Wade disappeared when Shumpert blew out his ACL and meniscus in a Game 1 blowout loss. Thanks to Amar'e Stoudemire's lack of defensive prowess, Chandler was forced to guard Bosh, a gambit that time and again forced the eventual Defensive Player of the Year away from the rim, leaving wide-open driving lanes for the vaunted Heat attack. Smelling blood, Miami exploited the Knicks' lack of depth and match-up disadvantages, torching them to the tune of 108.2 points per 100 possessions, making New York's the worst defense in the playoffs.

Still, when it was all said and done, the 2011-12 campaign remained the most successful defensive season for the Knicks in a decade, one that New Yorkers—especially the ones like myself who grew up watching the Riley and Van Gundy era Knicks—could be exceedingly proud of. The team had again forged an identity as a tough, blue-collar, hard-working, ass-busting squad that took no quarter, asked for none, dug in its heels and made you work your damnedest for every single basket. The remarkable turnaround happened mostly on the shoulders of the new guys, but old friends like Jared Jeffries and Landry Fields contributed heavily at different points as well.

67 Ironically, it was during this stretch that Carmelo Anthony played some of the best defense of his career–first at small forward, and then after Stoudemire was injured, at power forward, from which he put up some of the best, most efficient numbers of his career.

Jeffries, never known for his offensive game, probably could have led the league in charges drawn, had he been able to avoid injuries. Fields, though he often struggled to chase shooting guards around screens and guard bigger small forwards in the post, did an admirable job for large stretches of the season with his on-ball defense and hawking of the passing lanes. Even J.R. Smith, known for years as one of the laziest defenders in all of basketball, was able to find a place for himself in the lineup despite the worst shooting season of his career through hard work and perseverance in the defense and rebounding departments.

In the end, though, it all comes back to Tyson and Shump. When Shump was healthy, the two formed one of the most fearsome defensive duos in the league, with Shump hounding the ball on the perimeter and Tyson lurking behind him to thwart anyone that somehow managed to get by (which really only happened when Shumpert was navigating pick-and-rolls or off-ball screens; dealing with picks was his lone defensive weakness). Shumpert was masterful at guarding isolation plays, allowing scores on just 33.9 percent of the 115 ISOs he faced, while forcing turnovers 24.3 percent of the time. He guarded everyone from Dwyane Wade to Derrick Rose to Ricky Rubio, picking all their pockets and shutting down plenty of their drives.

By season's end, Shump had finished tenth in the league in steals at 1.7 per game. Of the nine players ahead of him, only Memphis Grizzlies defensive specialist (and noted Twitter celebrity) Tony Allen[68] played fewer minutes per game. All the while, he established himself as one of the best one-on-one defenders in the league, routinely earning TV praise from coaches and analysts alike for his defensive prowess—a combination of elite athleticism, quick feet and long arms that made it exceedingly difficult for opponents to get by him. And if they did? Well, there was Tyson Chandler.

68 Allen on Twitter during a pedicure: "Yo they got my feet in a sandwich bag full of hot lotion! But wait mis lady turn this elevator music off and let me play this ipad!! Cool"

FOURTH QUARTER:
MARCH 16—MAY 9, 2012

So, D'Antoni fled town and, in doing so, dumped his fizzling mess of a team into the lap of his fuzzy right hand man. The cooling embers of Linsanity, the fickle Anthony-Stoudemire-Chandler dynamic, and the narrowing odds of playoff contention had all become Mike Woodson's personal hurt locker.

Woodson inherited a team not only farting fumes but also one that would suffer massive injuries to key players at crucial junctures. Amar'e's back would shatter, then, once he returned, so would his self-control and one unlucky fire extinguisher casing. Lin's knee would unravel, ending his season and (little did we know it at the time) his Knick career. By season's end, Iman Shumpert and Baron Davis would fall as well.

The Knicks were 18-24 when Woodson took the reins and, with the coaching change and all those injuries, had every excuse to plunge into post-D'Antoni doldrums. But these wouldn't be the 2011-2012 Knicks if they didn't blithely disregard the most obvious of expectations. Practically overnight, Woodson stirred his team into a defensive club, a real unit, not just a bunch of guys and Tyson Chandler. Carmelo Anthony, outside the treacherous confines of the previous coach's widely beloved free-flowing offense, suddenly unearthed his jump shot and summoned the will to defend. When his teammates went down, he grinningly accepted more shots and hit a goddamn lot of them.

New York clambered and clawed through the rest of the regular season without losing consecutive games. They wrested pivotal wins from encroaching lower seeds. By the grace of Woody and Melo, they entered into the playoffs as a seventh seed, striding heads high and fists forward into...the Miami Heat. Shit.

Amar'e's take on the era: "Phenomenal."

A psychiatrist's diagnosis of Knicks fans: "Shumpotraumatophobia"

#1 single in Finland: Jukka Poika, "Ala Tyri Nyt"

Toney Douglas's Google search history: "toney douglas 2010 highlights," "nba players allowed to eat pudding on bench"

From the papers: "I really don't know what the situation is with that ... I can't really answer any of those questions about what happened, what he did or anything like that." (Carmelo Anthony on Amar'e Stoudemire's fistful of glass, *The New York Times,* 4/30/12)

Key stats: Down the stretch, from March 26 onward, Iman Shumpert averaged 33.2 minutes per game. It wasn't for his offense, because in 8 of those 17 games, he failed to score 10 points. Defense is in the hizzy.[69]

69 Immediately after typing this sentence, Mike Kurylo showed himself out. But not without copyrighting the door first.

CHAPTER 11
MELODIOUS

BY MIKE KURYLO
KnickerBlogger.net

What? Ten chapters without one on Carmelo Anthony? The Knicks' one and only true superstar? He's a collegiate champion. He's a gold medal Olympian. He can score. He can rebound. He can ... uh ... score. Did I mention he can score? Like a lot of points in a single game? If you cover the Knicks in a fashion that involves fan feedback, like us bloggers do, then you'll know how many fans have Melo-fever. Failing to indulge this faction will lead you to be called many nasty terms, most gentle of them being "a hater." But it's not like there were a lot of Melo tales to go around. He'd go 28 straight games without scoring at least 30 points. Scorers get the headlines because they make the nets (and the fans' hearts) go swoosh. No swoosh = no ink. And that was Anthony's 2012 season. Until now.

netw3rk
@netw3rk
Follow

"I'll show you who plays defense!" *karate chop*
-- Melo

Reply Retweet Favorite

When Jeremy Lin played his last game of the season the night of March 24, the Knicks were a near-even 24-25, good for the last playoff spot but still with a quarter of the season left to play. Ironically, the team's next game would pit them against the Milwaukee Bucks, who at that point were within a game of the Knicks for the eighth and final playoff spot. A Milwaukee win would put the two teams into a tie. Given that the Bucks held the tiebreaker, a loss meant the Knicks would be on the outside looking in.

Lin's injury, the severity of which wouldn't be disclosed for weeks, had left a gaping hole in the Knicks lineup. Indeed, Lin had been productive in filling the void at the team's weakest position—point guard. The *New York Post* declared Linsanity dead on their March 14 back cover, yet in the six games after that proclamation, Lin averaged

NEW YORK
7

14.5 points, 5.3 assists, and a true shooting percentage[70] of 58.8. Minus the services of their surprise savior, the Knicks not only lacked one of their main contributors; they lacked the best player at a position that was otherwise wire thin.

On the morning of March 26, Jeremy Lin and Amar'e Stoudemire claimed they would be suiting up for the all-important showdown in Milwaukee. By game time, however, both players were conspicuously absent. Behind Mike Dunleavy's 24-point first half explosion[71], the Bucks led by five at halftime. Undaunted, Carmelo Anthony racked up 10 in the third quarter—en route to a team high 28 on the night—and the Knicks went on to eke out a clutch 89-80 victory. Not only had the Knicks given themselves some breathing room against their division rivals; it was also the first time in two weeks that Anthony had managed to drop north of 20, and the first time in more than two months that he'd hit at least 28.

With Stoudemire's and Lin's health predictions ringing hollow,[72] Melo, now fully unencumbered by the disparate shadows of Amar'e and Lin, came into his own. Anthony would follow up his Milwaukee performance with 25 points in a win against Orlando, and 36 in a loss in Atlanta a few days later. Seven more times in the regular season he would score more than 30 points. From the time of Lin's season-ending shelving to the end of the season, Anthony averaged 29.1 points, 7.3 'bounds, and a true shooting percentage[73] of 58.5 percent.

In the playoffs, the Knicks were overmatched, outmuscled, and outhustled by eventual champions Miami. After putting themselves into a 2-0 hole, and watching Amar'e put a hole in a glass case—an outburst that would cost him and his team Game 3 and render the power forward one-handed the rest of the series[74]—Anthony responded. In game 4, on the verge of being swept for the second year in a row, New York found themselves down one with six minutes to go. Anthony, shoulders understandably quaking beneath a franchise's weight, scored 12 points down the stretch, including a clutch three pointer in the final minute, to put the Knicks up for good.

In the final five weeks of the season and into the playoffs, Carmelo Anthony owned the Knicks. Really though, his ownership of the team began before he even arrived in New York. In the summer of 2010, Chris Paul made a toast at Melo's wedding in which

---

70 Seriously, stop looking here every time true shooting percentage comes up. Recap—TS percent is an improved field goal percentage including threes and free throws. 54 percent is average. Got it? Good. Now go make me an omelet.

71 It's funny that this is footnote number 71, because we all thought Dunleavy was going to score 71 points.

72 Amar'e Stoudemire's balky back prevented him from playing for nearly a month, while Lin would find himself under the surgical knife just days later.

73 True shooting percentage is a made-up stat used solely for us statisticians to feel superior to other people.

74 Today STAT claims he didn't punch the glass-encased fire extinguisher so much as "slap it," because that makes the fact that his hand almost fell off totally okay.

he hoped aloud—the specter of the Heat's recently formed triptych fresh in the minds of all—that he and Carmelo Anthony would join Amar'e Stoudemire in the orange and blue. From then on, speculation that Anthony would join the Knicks climbed to a fevered pitch, with analysts screaming all the while that Gotham needed to do whatever it took to acquire the then Denver forward.

New Yorkers were suddenly being held hostage, as much by the media pundits as by Melo, who did his best to assure that he and he alone would benefit the most from any midseason deal. What happened next remains, in the eyes of many, a turning point for the franchise, as the Knicks emptied the kitchen cupboards in exchange for an eminently talented, but equally troubled, prodigal son. On February 22, 2011, it became official—Carmelo Anthony would join Amar'e Stoudemire in New York, which now touted dual All Star forwards expected to propel the Knicks into the upper echelon of the NBA.

Truth be told, of the two it had been Stoudemire who had failed to meet expectations. In Phoenix, STAT was one of the league's premier high-efficiency scorers, regularly boasting a true shooting percentage above 60 percent, including a league-leading 65.6 percent in 2008. His first year in New York, Amar'e's true shooting dropped a full ten percentage points (56.5 percent) from that high water mark. Speculations as to *why* continue to vary—was the drop in his performance due to his surgically repaired knee? His back injury? The extra weight he put on during the summer? Was Stoudemire simply a product of Steve Nash's panopticon greatness? Or—cue conventional wisdom—was it instead his inherent inability to mesh with Anthony?

Still, the "why" wasn't as pressing a question when it came to Stoudemire, if for no other reason than his narrative remained one that New Yorkers could get behind. After all, Amar'e had come to New York a full year and a half earlier than Melo, at a time when *no one* wanted to come to New York—not LeBron, not Dwyane Wade, hell, not even Joe Johnson.[75] By coming to a down-and-out franchise tortured by decades of close calls and doldrums, Amar'e had, for Knick fans, done something noble, although in reality his max-level deal wasn't much of a monetary sacrifice. Still, even to the most Dolan-hardened Knicks cynics, Amar'e was saving the franchise from its own disastrous history, wiping it clean so that other stars could potentially join him, and bring about a new Garden golden age.

Compare that to Melo, for whom the Knicks had paid dearly not only in dollars but in resources as well. New York is nothing if not a cauldron of consumerism, and its residents know a rip-off when they see one. So when New Yorkers were told by the media that Carmelo was worth more than all the assets they had given up, a bona fide top 10 talent, they expected to receive exactly that much in return.

75 <3 u Brooklyn

They may have been teammates, but Melo was never interested in comparing himself to Amar'e. His barometers were LeBron, Wade, and Durant, the game's Grade A-certified elite. And don't think for a second that Melo didn't recognize right off the bat the double standards at play between him and his frontcourt counterpart. Stoudemire was expected merely to bring the Knicks back to respectability. Anthony was expected to bring home the gold.

During Linsanity, the Knicks went a hardy 16-10. Not surprisingly, the media tripped over itself in turning Lin into an overnight sensation, contributing to an uber-cultish rise sure to be looked back upon in a manner not unlike that of a Raul Ibanez, Bucky Dent, or David Tyree. Still, it was hard to ignore the fact that the Knicks had gone an even more robust 12-5 down the Lin-less stretch, thanks largely to Anthony's Herculean efforts. Yet there was no Melosanity; no phrase-riddled tee shirts or hats or magazine covers; no globe-spinning hype. To this day, and despite his amazing run in the tail end of last season, the most lasting impression Carmelo Anthony has left on New York remains the drawn-out circus surrounding his acquisition.

From the moment the Knicks were identified as Carmelo Anthony's team, the Bronx-born prodigy inherited all the attributes of the franchise. Twelve men may have suited up in Knick jerseys every night, but only one would ever be truly to blame, even if his bespectacled, struggling teammate was making essentially the same amount of money; even if Tyson Chandler was more valuable to the team; even if the front office couldn't find two suitable starting guards. The team failed to win a playoff game in 2011, and that was Carmelo's fault. The Knicks started off 8-15, also Melo's fault. When D'Antoni was fired? Definitely Carmelo's fault.

Anthony has received, and will by all accounts continue to receive, the brunt of any Knick-aimed animosity. But how much of it is truly deserved? The NBA is rife with examples of general managers champing at the bit to overpay for scorers. So why is it the fault of Anthony—or Rashard Lewis or Gilbert Arenas or anyone else—for taking a max deal? Anyone reading this who says they wouldn't gladly accept a contract valued at one hundred million if one were placed in front of them is either lying, or insane, or both.[76] Carmelo Anthony didn't magically erase his defensive liabilities from game tapes when the Knicks were scouting him. He didn't force the Knicks to give up fan favorites Gallinari, Chandler, Felton, and Mozgov to acquire him. The Knicks made that move because they saw as many opportunities for banners as banknotes, and not necessarily in that order.

In reality, the frustrations and resentment perpetually laid at Anthony's feet belong to others. Blame the NBA for creating the system that overpays one-dimensional scorers. Blame the Knicks front office for overvaluing anyone that can drop in 20 points a night,

---

76 Rich people, take note: I AM ONE OF THE WILLING.

no matter what else they do. Blame James Dolan for signing the mortgage that gave away the team's future prospects. Like a lightning rod, Anthony absorbs the backlash for all these things, deserved or not.

But on the court, like it or not, Anthony played well in the tail end of 2012. When Stoudemire's back kept him from doing anything on a basketball court, Carmelo dumped 43 and 42 points against two of the league's best defenses in Chicago and Miami. When Jeremy Lin was resting his knee, Melo dropped 35 points against the Celtics and followed that up the next night with 33 against the Nets, both division rivals.

For Anthony, a handful of regular season wins won't get people to pay him some proper respect. The first playoff win in 13 years won't bring adoration. One series or even two isn't going to cut it. Carmelo Anthony's Knicks will likely need to reach the Finals before the fans feel like they've gotten their money's worth. And for some even that will never be enough.

DON'T FORGET ABOUT RENALDO BALKMAN

A.K.A. HUMPTY[77]

BY SETH ROSENTHAL

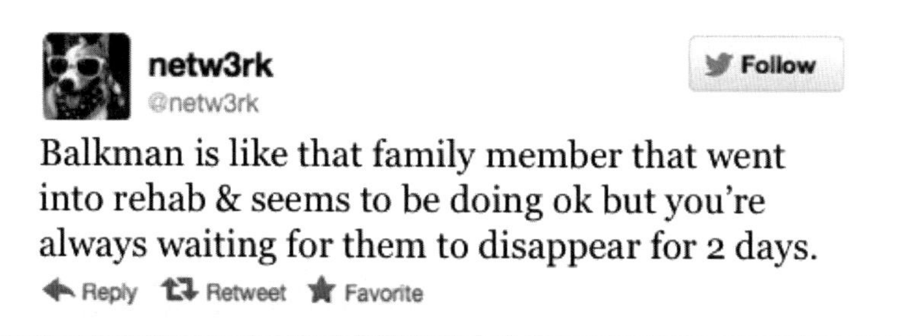
netw3rk
@netw3rk
Follow

Balkman is like that family member that went into rehab & seems to be doing ok but you're always waiting for them to disappear for 2 days.

Reply Retweet Favorite

Renaldo Balkman was a retread of pure happenstance. He was the best kind of retread—one with bristly dreadlocks pouring out the top of his headband and a lightly medicated expression—but an insignificant, inadvertent retread nonetheless. After being unceremoniously traded several years prior (something that upset his cult fans, but probably not him even a little bit), he'd floated back to New York in the Carmelo Anthony trade the previous season, spending the remainder of that season snoozing on the bench.

The organization regarded Balkman as more of a moveable contract than a playable asset—only Isiah Thomas, the man who drafted him, ever saw much utility in Balkman's somnambulant style of play. There was a brief period, though, during the abbreviated preseason, when Balkman looked like he might crack the rotation. He was at his dazed, reckless best in a pair of games against the Nets, cruising away from the ball to fill open lanes, finish, and give the occasional delirious hard foul.

It wasn't enough, though. Balkman played just garbage minutes to open the season, only to be cut to make room for the J.R. Smith signing. One wonders if Renaldo might've gotten his chance had he lasted until the similarly sleepy and defensive-minded Mike Woodson took over, but it wasn't meant to be. Balkman, to his credit, didn't really seem to mind. He sidled out of New York in the same unheralded way he returned. If you were looking for him after that, you could find him on Twitter, posting photos of himself rollerblading, or in Puerto Rico quietly preparing to play for the hapless national team that adopted him. Renaldo never really missed a beat.

77 Hi everyone!

CHAPTER 12
THE BEGINNING OF THE END

BY JAKE APPLEMAN
The New York Times

By late March, the initial glow of Linsanity had worn off—we were almost used to an undrafted Asian American starting point guard from Harvard running the show at the World's Most Famous Arena—and the Knicks were looking to fully stake their claim on a playoff spot. Other than the dominance shown by Miami and Chicago, the Eastern Conference was a jumbled mess. Woodsanity had taken over. The Knicks won five in a row since Mike Woodson had stepped in for Mike D'Antoni, who stepped down. This story begins in Los Angeles on March 23, the same night the Knicks had their winning streak snapped in Toronto. If someone had told Mr. Appleman that he would be watching Jeremy Lin play his final minutes as a member of the Knicks the following night in New York, his reaction would have been something out of a daytime soap opera. But nobody knew anything, nobody said anything, and everybody yawned.

When the beginning is like nothing you've ever seen, the beginning of an unexpected end can go unnoticed, like a comet streaking across the sky while you trudge along, head down, listening to crickets. It's surreal when the man who was the world's lede only a few weeks ago can barely make the ninth paragraph of your game story in what would be his on-court swan song.

Despite the fact that he went unmentioned in Woody's pregame presser after the win against the Pistons, Linsanity continued apace after #17 took his final steps in a sweat-soaked game jersey towards the bench with 4:47 remaining in the third quarter, the Knicks up 10 against a lottery-bound team.

There was practice the next day during which Lin beat the term "game time decision" into the ground. There was the eighth seed battle against the Bucks he sat out, won by the Knicks, a game made all that more enjoyable by the postgame presence of the reigning Eastern Conference player of the week, Drew Gooden.[78] And there was eventually Traysanity, the day Tracy Morgan showed up at Knicks practice to entertain the overworked scribes with his unmistakable brand of humor.

The night after the victory at home against the Bucks, Lin was trotted out in front of an eager crowd at an event sponsored by Steiner Sports Memorabilia and *The New York Times* store, a unit of the Times Company that sells memorabilia and other items. Steve Novak and Iman Shumpert later joined him on stage. Lin was never going to play again for the Knicks, but opportunistic New Yorkers were ready to hang on his every word.

I was in the green room with Lin, his family, his then-agent, Roger Montgomery, and Brandon Steiner; the evening's MC. Montgomery became indignant when a tabloid reporter and I asked Jeremy about his knee. Eventually, we did interview him and he talked about Novak's unpredictable humor and his love of collecting almost exclusively Michael Jordan cards as a kid. (The reference to my Penny Hardaway collection losing value over the years fell on deaf ears.)

Lin went out on stage to greet a different kind of adoring crowd. I hung back with Novak and Shumpert in a side room, interviewing Novak and watching him chirp at Shump for cheese discrimination. (Novak is from Wisconsin; he demands that you approach a fruit and cheese plate with the same discipline of a sharpshooter rotating around the arc during warm-ups.)

The event was the most relaxed I had seen members of 2011-12 New York Knicks at any point during the season. The start of the season was misery. Linsanity was a purified whirlwind. And Woodsanity was nice because the team was winning, but not an altered mind state or new reality necessarily worthy of a defining moniker.

Now these Knicks, with a day off from practice, were attempting to breathe. And, as is the case when three unselfish, underpaid underdogs exhale together—the results amplify their crowd-pleasing reputations to the nth degree.

Lin was far more engaging on that stage than he ever was with the New York and national media on a day-to-day basis. He honestly discussed his wish for greater privacy and the compromise that came with exploding into the world's consciousness while he was beginning to achieve his dream. He wasn't particularly keen on avoiding phrasing that could be misinterpreted, just saying what he wanted to say and making sure he got it all in.

He was selling his brand, but more than going along with what everyone else wanted from him, he was doing it on his own terms (or his mom's terms) with a company that would become one of his few, selected corporate sponsors.

Steiner brought up the poll from ESPN that ostensibly showed Lin to be the most

78 Somewhere, the president of the Drew Gooden Fan Club just declared himself Dictator for Life.

popular athlete in New York. Lin cited social media as one obvious reason for his soaring popularity.

"I'll get tweets from different languages I don't understand," he said. "I obviously assume they are nice things."

People laughed because Lin subtly acknowledged that all the widespread praise was sprinkled with people digitally ripping him in tongues ranging from Swiss-German to Swahili because that's the way life works in 2012.

Lin openly credited Jeffries for embracing him before he ever played a minute for the Knicks and Tyson Chandler for showing him the ropes. He joked that he played better in front of his mom because he didn't want to have her in his ear afterward.[79] "I used to try not to look up at her, especially after a turnover," he said. More laughter and a round of applause. It was not a tough crowd to please, but good stuff to witness if you're capable of not being cynical about everything.

Shumpert and Novak joined him on stage. Steiner immediately ribbed Novak, referencing Lin's joke that Novak only runs from one three-point line to the other. Novak got Lin back, playfully scolding him for not mentioning the one game they played together with the Reno Bighorns when Lin went over the teammates who helped him when he first arrived with the Knicks.

"I'm sorry, when I [said] Tyson and Jared gave me a lot of advice, I actually meant Shump and Novak," Lin responded with confidence.

Jeremy Lin, aware of his surroundings, slinging zingers. Who knew?

Shumpert furthered the likability theme, discussing growing up playing a rougher version of basketball on the streets of Chicago, and embracing the role of defensive stopper, at times because he simply didn't like his opponents.

"I hate their whole swagger," he said of some opponents. "You know what I mean? 'I just don't like you.'"

"Novak takes a different approach," Lin said. "He lets the guys score."

The laughter eventually died down. Lin kept messing with Novak's defense.

"You get a three. I get a three. You get a three. I get a three."

That segued into a discount triple check discussion and then Novak's Milwaukee upbringing[80] and certified Cheesehead status. Novak immediately pointed out that Shumpert, a Bears fan, was eating cheese before coming out on stage, the cheese discrimination moment coming full circle.

The conversation continued with laughter and jokes bracketing revelations and repetition. Shumpert closed it out in style by rapping for all in attendance. His performance was

---

79 Seriously, how bad would it hurt to have your mom crawl inside of your ear?

80 Fact: If you order a garden salad in Milwaukee, they bring you a plate of meat.

impressive, at least by NBA player standards, and the players received a rousing ovation.

The night was a testament to close-up, well-moderated publicity, as some of the Linsanity fairy dust had obviously rubbed off on both Shump and Novak. Novak used the analogy of Lin's "wake," as if Linsanity was a cresting wave that he and others were able to surf.

"When the Garden gets loud, it's the most unbelievable thing I've ever experienced in terms of being in an arena, playing in front of fans," Lin said.

The producers have decreed that he will have no orange-and-blue encore.

CHAPTER 13
WHERE DO WE GO FROM HERE?
FEATURING JARED DUBIN, DAN LITVIN, AND JAMIE O'GRADY

This email chain started a few days after Mike D'Antoni left the Knicks in the midst of another long losing streak. The last email was sent just minutes before the Knicks beat the Raptors soundly for the fourth consecutive win. Winning streaks like the one the Knicks rattled off in the early days of Woodsanity can often make you think all of your problems have been solved. Obviously, we knew better. This Hardwood Paroxysm *roundtable first appeared in late March 2012.*

Dubin: Well, it was a quiet week in Knicks land. Not really too much to talk about. I have no idea how we're going to make an article out of us emailing each other with all this boringness and lack of activity. Oh wait; this week was absolutely insane, even by Knicks standards. My B.

We'll start, obviously, with the jettisoning of head coach Mike D'Antoni. Where do you guys stand on this? Is it the result of New York's almost maniacal savior complex? Was it a case of a coach who should have been able to make things work with Carmelo Anthony and instead let his stubborn refusal to alter his system to suit the talent on his team be the end of him? Was he a coach set up for failure and then wrongly held accountable for doing exactly that? How do you feel about his Knicks tenure as a whole?

Litvin: This is the first I'm hearing about this! WHAT!??! In all seriousness, I've always been a big Mike D'Antoni fan, and I was upset that he left. That said, D'Antoni was definitely set up for failure. The Knicks lured him to New York with the promise of an embarrassment of riches in terms of talent if he'd suffer through two years of Nate Robinson and Chris Duhon.[81] Well, that and actual riches. As he sullied his coaching record leading a revolving door cast of goons and talentless hacks, the press and fans talked about the so-called free pass Donnie Walsh had given him, but always in the context of why he was a crappy coach that couldn't maximize the "talent" he was provided. Talent, like Larry Hughes and Darko.[82] From day one it seemed as though these observers would hold D'Antoni to the exacting standards to which it

81 We're convinced that the "Once a Knick, Always a Knick" stitching on the inner neck piping of the new Knicks jerseys was meant for these two and these two alone.

82 Milicic, as in the guy who was drafted immediately after LeBron and before both Melo and Dwyane Wade in the loaded 2003 NBA Draft. Somewhere, Joe Dumars kicks a unicorn right in the balls.

holds every coach, despite the common knowledge that we were all supposed to just be biding our time for LeBron James. And the "FIRE-D'AN-TO-NI" chants indeed did parrot down from the rafters in D'Antoni's first year.

But intelligent people knew that he would never be terminated unless he failed *after* the Knicks provided him with some, you know, *basketball players*. Donnie Walsh did exactly that in the summer of 2010. It wasn't the all-out coup that a LeBron James would've netted, but Amar'e Stoudemire and Raymond Felton blended nicely with some of the pre-existing young talent like Danilo Gallinari and Wilson Chandler and the Knicks were seen as a rising young squad that was playing with confidence.

But then Carmelo came along. He knew he could join the Knicks as a free agent in the summer of 2011 but—knowing that the inevitable upcoming lockout would likely mean a reduced payday—knew he needed to be *paid*. So he and his people convinced James Dolan to gut the team for him. Thus, shortly after it seemed like the Knicks were living up to their promise to D'Antoni by giving him a chance to succeed, they knee-capped him by getting rid of his favorite players and replacing them with a guy who fans believed could have made it work with D'Antoni but probably should have known better.

Carmelo just didn't fit—or hasn't yet, anyway. Or maybe, D'Antoni just didn't fit anymore. I guess it doesn't matter. Neither showed any inclination to change, but one of them had to be sacrificed to Moloch. Of course, it's always easier and quicker to jettison the coach. That's just the nature of 21st century sports and sports stardom. But the *way* it went down left a bad taste in a lot of people's mouths, mine included, not least of all because Carmelo's arrival twice led to upheaval of good vibes and pretty basketball.[83]

But really, why couldn't it work with Carmelo? I mean, what was Mike D'Antoni asking Carmelo to do that was so awful or out of the realm of possibility? Move the ball? Move without the ball? Space the floor? Stop impinging on the point guard's space? For whatever reason, Carmelo just didn't feel like doing those things. That contributed heavily to the fact that the D'Antoni Knicks were 25-38 (including playoffs) in 2010-11 and 2011-12 with Melo, and 35-30 in games without him.

Clearly I'm using this question as an excuse for catharsis, and that's why this answer is so long-winded, so I'll just close out by saying that I'm not trying to completely absolve D'Antoni. Could he have done something different? Absolutely. He could've bent to Carmelo's will and just played how Carmelo wanted to play. He could've done that and won more games with Carmelo than he did, bided his time, and bounced at the end of the year at the expiry of his contract. But D'Antoni believes the

83 Once after the trade, and the other when he [arguably] throttled Linsanity.

game should be played a certain way and he stuck to his principles instead of giving in to a pampered star's petulance.

O'Grady: Ugh. I'm already on record as being mostly disgusted with the premature cessation of D'Antoni's Knick tenure, but the team's play since his departure has only reinforced the fact that the relationship had devolved into a marriage made in an NBA Hades, at best. Hades, as you know, is the land of .500 records, a purgatorial weigh station of sorts, where you're not good enough to win a championship, nor are you bad enough to obtain an elite player in the draft.[84] In other words, almost every Knicks season since "the frozen envelope" brought Patrick Ewing to New York.

But I digress...

In the end, D'Antoni was part victim of circumstance and part slave to his own stubborn philosophies. And, unfortunately for him, Gotham is the one place where time is never an ally. Whether the stache-lovin coach got what he deserved or was simply dealt a bad hand certainly remains open to debate, but it is impossible to deny that rare were the opportunities for D'Antoni to work with a cohesive, drama-free roster for any extended period of time.

In fact, other than the first half of the 2011 season, when the team was starting to coalesce around the pace and style of play favored by D'Antoni, at no time was the former Knick coach able to do his job under even a modicum of normalcy. Taken together, it seems unfair—to me, at least—to judge D'Antoni based almost solely on three and a half years at the helm of a mostly rudderless ship, and a one-month stretch in which the team's "superstar" small forward had taken it upon himself to submarine his head coach. Ironically, though D'Antoni was ultimately held accountable, one of his principal failures was that he never held his players *publicly* accountable. In this town, it's all about perception. Perhaps if D'Antoni had called Carmelo Anthony out in the media for his unwillingness to "play ball," so to speak, the coach might have established some cover for himself.

Either way, D'Antoni is gone, and now the players have no more excuses.

Dubin: This particular bit of Dan's response interested me:

He could've bent to Carmelo's will and just played how Carmelo wanted to play. He could've done that and won more games with Carmelo than he did, bided his time, and bounced at the end of the year at the expiry of his contract. But D'Antoni believes the game should be played a certain way and he stuck to his principles instead of giving in to a pampered star's petulance.

Dan, I have to wonder, do you think he should have given in for the sake of win-

84 Actually, Jamie, Hades was the name of the Greek god of the underworld and only later came to designate a place where dead people and Isiah Thomas lived.

ning games rather than sticking to his principles? If he had, would he still be here? Or was this—teaming D'Antoni, Anthony and, to a lesser extent, Stoudemire—a doomed experiment from the start?

Litvin: Well, maybe it's a distinction between *could* and *would*. I think he could, but he seems very principled—too principled to change for the sake of one player, even. I guess one could argue that this makes D'Antoni a poor coach, for lack of flexibility. But D'Antoni probably wants his identity to be associated with what we saw during Linsanity, and I bet that's part of the reason he felt comfortable quitting. Applying for his next job, he can declare, "Look what I can do if you give me the right pieces." On the other hand, if he sacrificed his system and led the Knicks to a good record with Carmelo, that would have probably looked even better on his resume.

I suppose it does beg the question "was teaming these guys a doomed experiment from the start?" Two rigid, headstrong principals vying for supremacy clearly was not good for the locker room. And it was probably misguided for fans to ever think that Carmelo would be willing to change for a coach that he clearly did not respect.

O'Grady: The question of whether D'Antoni's doom had more to do with stubbornness or sheer roster construction is not an easy one to answer. Let's break it down separately.

The conventional wisdom that D'Antoni was unapologetically hard-headed in refusing to bend his will to the "needs" of his star players is based mostly on supposition, not fact. I can confirm that Mike and his staff, even before the addition of Mike Woodson, preached both defense and rebounding. When Melo was first acquired last February, D'Antoni made it a point to focus his efforts on the defensive end of the floor, and though the offense was a work in progress, we heard nothing about the system not being viable. Only when the losses piled up in January did whispers of Carmelo's dissatisfaction begin to surface. And yes, we all saw the point forward experiment[85] fail miserably with Melo earlier this season, but can we at least acknowledge that Melo himself is on record as saying he likes the offense run through him? Tough to run it through him any more often than the Knicks tried to for the first month of the season, though I realize being a distributor is—le gasp—different than being a scorer.

On the possibility of STAT 'n Melo making things work, the statistics thus far are pretty conclusive—and not in a good way.[86] Notwithstanding the last four games—all Knick wins—Stoudemire and Anthony may very well be a poor fit together. That they don't play a lick of defense certainly doesn't help matters, either. Still, I would argue

85 During the early part of the 2012 season, Mike D'Antoni used Carmelo Anthony in a role similar to that of LeBron James, letting his star forward orchestrate the offense from the perimeter and elbow. Toney Douglas was fine with this.

86 Zach Lowe did some fine work on this very subject for *Sports Illustrated*.

that any offensive system not run by Phil Jackson needs a serviceable point guard to function, and since we only got to see Jeremy Lin play with both Melo and Amar'e for a brief stretch of time, we will never know what the group could have achieved under MDA.

The sample size under Woodson is small, but the effort on display is anything but. Is it disturbing that professional athletes, paid tens of millions of dollars, were apparently not exerting maximum energy for their previous coach? Sure. Surprising? Not really.

Dubin: We're going to go deeper in Carmelo,[87] but just to finish up with him and D'Antoni here, let me ask a few more questions: Even if Carmelo didn't change his game, do you feel that he could have been successful in D'Antoni's system if he bought in entirely? After coming over in the trade from Denver last season, he played arguably the best stretch of basketball of his NBA career (his 22.8 PER in those 27 games would have represented a career high). Is there any particular reason you think that success didn't carry over to this year? Was it the always-doomed "point forward" experiment? Sheer stubbornness? Did Stoudemire's poor play just mean that too much defensive attention was paid to Melo and he couldn't find his offense as easily (remember, Melo and STAT were the highest scoring duo in the league after the trade last season)? In short, *what happened?*

O'Grady: I have no doubt that if Carmelo had bought into D'Antoni's system, he would've met or exceeded his career scoring output, and the Knicks would've won a lot more games as a result. Has there ever been a player not named Starbury[88] who has played under D'Antoni that didn't see his offensive production increase? Under Woodson thus far, the Knicks have basically run the same motion offense as they did under MDA, with a few more isolation and post-up plays for Melo sprinkled in. Maybe that's all Anthony ever wanted all along, but Melo is still shooting just as poorly under Woodson as he was under MDA. It's his all-around effort and energy that have improved, and those should have been there all along.

Litvin: I agree with Jamie on this one. What could have *really* changed so dramatically between around 10 or 11 a.m. when news broke that Mike D'Antoni would resign, and later that night when the Knicks manhandled the Portland Trailblazers? If all it took was a bunch of platitudes about accountability and defense to get this team to play hard then Mike D'Antoni was a miserable failure for not providing the requisite lip service.

87 That's really gross, Jared.

88 Coney Island native, headstrong journeyman point guard, and early 2000s Knick Dark Ages mascot Stephon Marbury, who once ate Vaseline on camera. YouTube it.

But I don't think that's what happened. Again, the team *did* play hard and give all out effort during Linsanity. And Melo was largely absent for that. When he returned, team effort vanished into the ether, and I think D'Antoni knew it wouldn't come back as long as Melo continued to play disinterested, aloof basketball, which he would continue to do so long as D'Antoni was his coach. That brand of half-assed ball rubbed off on teammates and affected them mentally, undermined their confidence, which is an attribute that has clearly returned to the team in spades. Clearly a weight has been lifted off of Melo's teammates' backs, largely because Melo—the team's captain, after all--has no excuses anymore for his failure to compete.

In the three games since D'Antoni left, what's been different about Melo? It certainly hasn't been his dreadful shooting, which has only continued. But we have seen Melo hustle for loose balls, be more aggressive on defense, and play more physically than we've seen him play as a Knick, period. If people want to think that's because Mike Woodson fed the press some sound bites, well, I don't know what to tell you. The reality is that Melo knows it's his butt on the line now.

Dubin: You've both mentioned Melo's increase in effort level since the coaching change, which is something that I've noticed too. It's definitely a good thing, but I also feel like that's not really enough for a lot of New York fans. I still get tweets before, during, and after games about how "Melo sucks, he's a bust and a bum and he needs to be traded and he'll never win." Do you feel like he has to score at a superstar level to win with fans in this town? Or can he get by being part of the overall team concept, hustling, making plays and scoring when he needs to? Basically, if he's not the Melo we thought we traded for, and the team is still good, will NY be okay with that? I think they (we) should be, but I get a hunch that plenty of people won't see it that way.

O'Grady: You bring up a great point and it all comes back to that infamous New York perception game. We've seen Big Apple fans be fine with "superstars" playing more like role players—the 90s Yankee teams come to mind—but we've also seen their ire directed at players whose individual performance doesn't translate to the team's overall success, as with A-Rod and even Ewing to some degree. Melo, like Big Pat and A-Rod before him, will ultimately be judged—and therefore embraced or derided—based on his ability to bring the team to the promised land. Anything short of a Garden-championship for Gotham on Anthony's watch will be met with scorn as intense (if not more so) as the hype was before his ballyhooed arrival.

Dubin: Do you think Melo is in the same type of situation that LeBron is,[89] though, where he doesn't just have to win, but be the most important and best player on the team and make all the big shots while winning or people won't really care? LeBron is

89 Was. Very much "was."

in such a precarious situation in Miami because he has Wade with him, and if he's not literally the one to carry the team to victory all by his lonesome, there will be many who feel he hasn't accomplished enough. I feel like people in New York are so starved for a Garden banner that it won't matter, but there's also such palpable Melo hate from some people here that I really don't know. Thoughts?

Litvin: I think if he can be a good player on a winning team, people should accept it. I could care less if he scores 30 points, as long as we win. I'm not convinced we need to waste bandwidth on this, though, since there's really been no indication that Melo is going to be anything other than the Knicks' primary scorer. It's true that in the last few games, the raw numbers suggest that Melo has sacrificed shots and points. But in the games since Woodson took over, which have all been blowout wins, Melo has played just 24, 22, and 29 minutes.

O'Grady: I am not sure LBJ's and Melo's situations are analogous in that the former's team is a championship contender regardless, while the latter is a sub .500 player with *his* "super team." The style v. substance debate may be irrelevant here if the wins don't come. This is actually the opposite of what most expect from Anthony, in that he's been pegged as a player whose "gets mines" trumps wins. Now he's losing and no longer getting his.

Litvin: I can't speak for the haters out there who won't be happy even if the Knicks win a title if Melo isn't the man on that hypothetical squad. That makes no sense to me. Is it possible that if the Knicks win the title and Melo isn't the hero, that people will question him the same way they do LeBron? Sure. People will say he's not like Jordan or Kobe or Tim Duncan. But that's not a valid baseline assumption anyway as far as I'm concerned. Personally, if the Knicks win a title I'll be permanently exhilarated to the point of incoherence. I won't be able to entertain such thoughts, let alone tie my shoelaces, or chew my food.

O'Grady: The Knicks will never win a title as long as I'm alive. Sorry, but I made a deal with the devil in 1996 in exchange for that Yankee ring. I feel horrible about this. My bad.

Dubin: Let's just go here now: Is Melo the most divisive Knicks player of your lifetime? You guys are a little bit older than I am, but I really can't remember another Knick drawing such a clear dividing line amongst fans. I've said this before, but in the last few weeks leading up to D'Antoni's exit, it felt like Knicks fans *had to* choose one of D'Antoni and Melo and defend them to the death. You couldn't get away with blaming things on both or people from each side would come at you. Can you remember another Knick causing this kind of reaction and dichotomy between fans?

Litvin: I don't know if Melo is the most divisive Knicks player of my lifetime or if it just feels that way because now we have blogs, but especially because we have Twitter. Imagine if Twitter was as popular as it is now when Marbury donned a Knicks uni, or if it existed in Ewing's heyday.[90]

My recollection is that the vast majority of Knicks fans (including myself) were thrilled when Isiah gave up so many assets to bring in Marbury. But somewhere along the line the perception turned, and during that process I remember there being division between the pro- and anti-Marbury people (obviously with the benefit of hindsight there is near unanimous agreement amongst Knicks fans that Stephon was the wrong guy to whom to hitch our wagon). You're right though: Even given some time to cool off, it's difficult to get folks to parse out blame other than exclusively to one or the other party. For example, I have followers on Twitter engaging in semantic gymnastics just attempting to interpret the following quote from Melo:

The last three games my focus was to have an energy that I haven't had so far this season, especially on the defensive end.

Look, it's clear that D'Antoni didn't motivate Melo to play hard but doesn't Melo also have a professional responsibility?

O'Grady: There was always a sharp divide amongst the fans and media alike when it came to Patrick Ewing. Ironically, he was a ball-stopper, too, and a stubborn one at that, but his defense, rebounding and willingness to play hurt alleviated much of the consternation. Patrick's teams also won because of his contributions, not despite his lack of others. So Melo will be viewed more critically until he consistently puts forth maximum effort on both ends of the floor every single night. We can live with failure from a player, but not indifference. Also, keep in mind we now live in the instant-access, 24/7 Twitter-sphere, where any schmuck with a blog, no offense intended, gets to speak his mind. It's just the way of the world now.

Litvin: "We can live with failure from a player, but not indifference."

Bingo. That's why Ewing will forever be a legend, while Melo still has *a lot* to prove. Also, Jamie, go easy on yourself. You're more than just a schmuck with a blog.[91]

Dubin: Speaking of failure and indifference, let's talk about Amar'e Stoudemire. Seemingly, he just gets a big old pass from both the New York media and a lot of Knicks fans. Whether this is because he was the first one to sign up for the cause or because they view Melo as a much bigger problem, I'm not sure. What I am sure of is that he's been just as bad, if not worse—okay, worse—than Melo this season, and it hasn't been nearly as big of a story. How worried about STAT are you going forward? Are you okay with him getting this "free pass" from fans and media? Is there anything

90 It did exist. It was called Prodigy. Sheesh.

91 No, he's not.

that can be done to bring *last year's* STAT back?

Litvin: I happen to think STAT has lost a step so, in my opinion, we can forget about ever seeing last year's STAT again. I mean, he got rejected by the rim in Indiana on Saturday. WHEN HAS AMAR'E EVER BEEN REJECTED BY THE RIM? It's sad, really. He has to adjust to be more effective below the rim, and I actually think we're seeing some of that now. He's still quick enough to make moves around the basket and the last few games we've seen him repeatedly employ a nifty little spinning reverse lay-in (he used to have a spinning reverse dunk). And I do think the shot will come back, and his numbers will improve.

Amar'e does get somewhat of a pass from fans because he "saved" the franchise from suffering the same fate as the 2000 Chicago Bulls team that employed a similar cap-clearing strategy to target Tracy McGrady and Tim Duncan and instead came away with Ron Mercer. He may also get somewhat of a pass because he's more skilled than Melo at saying the right things to the press, because he seems to be a genuine person, and because maybe, just maybe, he duped us into thinking that he's just incapable of playing defense (it's still not great but it's improved under Woodson), boxing out, and performing other, non-offense related duties, when in reality he may have just been neglecting to try.

Note, though, how he only gets "somewhat" of a pass. That's because I think fans are cognizant that Amar'e's been poor on offense and piss poor on defense. It just isn't talked about as much, owing to the variety of factors (above) enabling him to come across as likeable, combined with Melo's tendency to attract controversy.

O'Grady: The key difference between Amar'e and Melo is that Amar'e knows what to say and when to say it. Stoudemire, who ironically put his foot in his mouth repeatedly during his Phoenix tenure, is always accountable nowadays and rarely offers excuses for his poor play. Melo, on the other hand, has been something of a drama queen since he manipulated his trade to New York, and as such has a far shorter leash from the fans.

There's no denying that both players have disappointed with their play this season, but for whatever reason, Amar'e is viewed as a more sympathetic figure. It's worth noting that D'Antoni himself believed that the Suns would have won at least one or two rings if they had a power forward with even half of Stoudemire's basketball-IQ. He simply isn't an intuitive player, and no amount of coaching will ever fix his inadequacies. His supreme athleticism has carried him throughout his career until this season, but it remains to be seen whether he can grow from a cerebral standpoint as the physical tools further deteriorate.

CHAPTER 14
DYING IN LINTENSIVE CARE[92]

BY JIM CAVAN
KnickerBlogger.net

With the Knicks once again making an early playoff exit, disappointment quickly turned to excitement that the season's bottled lightning—Jeremy Lin—would soon be sealed up for good. By virtually all accounts, the Knicks were determined to retain Lin at any cost. But as the summer drew on, a noticeable lack of communication between the two camps lent those mired in fatalistic angst all the excuses they needed to start freaking the fuck out. What followed was a series of PR bungles and management fuckups so inexplicably awful, even Knick agnostics couldn't help marvel at the sheer shitshow unfolding before them. Indeed, if Linsanity had begun with a bang, it was about to leave with the weirdest of whimpers.

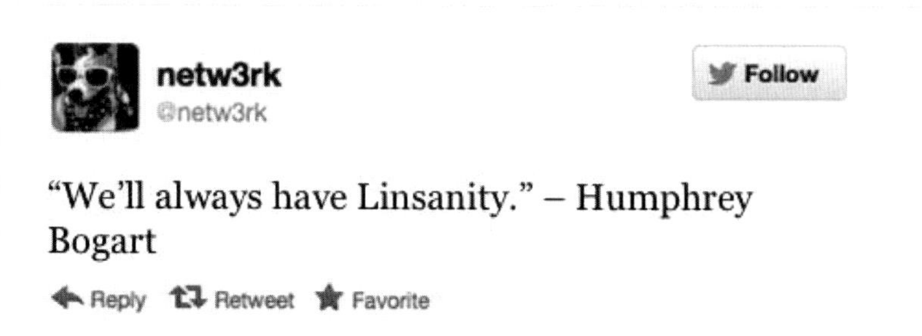

Like being stuck for days in a hospital waiting room—that's how awful the whole thing was. Tired and hunched over, sweaty, palms clammy, a steady shake and occasional twitch, you look up and down the hall every few minutes for word in the form of shadows cast by a long white coat. Every now and again you get up and pace down to the vending machine to count all the letters on all the bags and packages, just to keep your mind occupied, adding them up only to walk back empty-handed and already forgetting the number. The closest thing you've had to a shower is washing your face and hair with the horrible pink hand soap. Eyes so dark and baggy you squeeze off eye crust the size and texture of Chicken McNuggets.

The patient? Nothing more than the avatar for franchise rebirth, Jeremy Lin, a

92 He'll show himself out...

redeemer in the purest sense of the word, who'd rescued the Knicks' faltering campaign from a February doldrums fast approaching mutiny-level acrimony. Now, mere months later, that entire, blissful episode exists only in tiny YouTube pixels, and the mind's eye's occasional spark, his status having long since devolved from life support to simple death watch.

The condition: Attempted assassination, perpetrated by a franchise in the full throes of an acute brain rot brought on by blind pettiness, lack of foresight, and too many other maladies to name.

The prognosis: We'd already started carving the digital epitaph.

We'd have packed it all in—the clinging to last-second miracles, the online petitions,[93] all of it—if we hadn't appreciated that the kid had been on life support once before. Five months earlier, to be exact. With the season hinging on the sore limbs and groins of its two supposed superstars, Mike D'Antoni eyed the end of his bench and called, out of nothing more than sheer desperation, for the D-League-doomed kid wearing #17 beneath barely wrinkled warm-ups. Had the team's embattled skipper not made the move, chances are the cast-off Harvard grad—who'd already called three teams home, each for briefer and briefer spells, in his 16 or so months in the league—would've found himself once again cast atop the waiver wire chum bucket. Like a dying man who finds his last, best hope in the company of the similarly doomed, the Knicks reached as far as possible down the dark corridor, finding less a sliver of light than a brief respite from the dark.

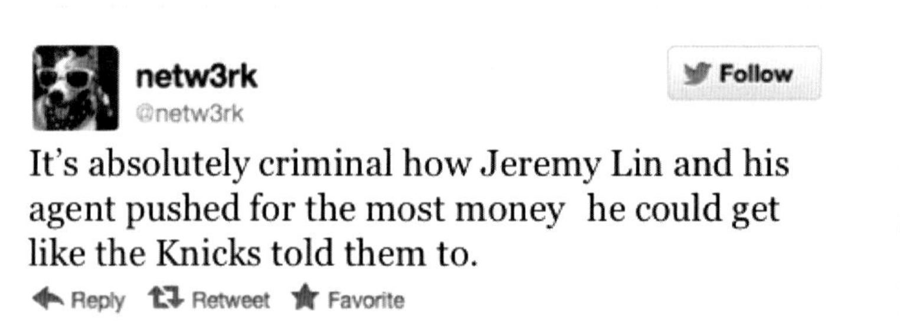

We all know what happens next, though 100 people would probably give you 100 different one-word superlatives. It was the strangest of sports phenomenon, in the sense that the whole thing seemed simultaneously to last for eternity and to end in an instant.

93 There really was a petition to keep Jeremy Lin in the orange and blue. We can't remember who actually started the thing, but Knick Knation, aided by the power of Twitter, was able to get north of 10,000 online signatures in less than 48 hours. IT TOTALLY MADE A DIFFERENCE.

Within weeks, jerseys, tee shirts, plush dolls, posters, magazine covers and ill-conceived poster boards alike propagated unfettered, as a feel-good sports story morphed into a bona fide global economic phenomenon. With every impossible performance, the Knicks—long cemented in the NBA universe to a kind of resident villainy—were suddenly becoming *likeable* again. Even after the gaudy numbers petered out and the wear and tear of perpetual fifth gear had ground down his knees, Lin looked a lock for orange and blue fatigues for at least the next decade. And a team that had spent even longer than that mired in steak-less sizzle and April vacations had uncovered its best chance at putting an end to a drought nearly four decades on. But, as with most things Knickerbocker, the water would wind up being another mirage.

* * *

At around 11:00 on the night of July 11, sourced confirmed what had been accepted as hideous gospel for the better part of two days: that the Knicks, going against every wink, nod and cap tip they'd made the month previous, would not match the Rockets' offer sheet. Lin was going to Houston, and with him the tails of a shooting star as liable to light the NBA world as peter out in dust—it'll be years before we know which. Just like that, quicker than a fleeing Midtown taxi, the vital signs finally, mercifully flattened. Linsanity was dead. As Knicks fans understood it, anyway.

Of course the show would go on. Even if you had come to live and die with what Linsanity meant to New York's peerlessly dysfunctional franchise, you couldn't help summon a modicum of guarded happiness that this most temporary of flings was headed where he'd surely be loved. After all, barely a year had passed since Houston's previous cornerstone, Yao Ming, had announced his too-early retirement,[94] leaving in his wake an international fan base clamoring for a fresh vessel. In Houston, Lin should be able to seamlessly transpose his international superstardom onto a ready-made apparatus eight-plus years in the making, and oodles of overseas appeal—in China, specifically—to its credit. Coupled with Houston having been the last team to cut Lin adrift, the full-circle nature of the deal seemed downright charming. That is, of course, until one comes back to the fact that the whole thing—from go, and at nearly every conceivable turn—needn't have gone down like it did.

It didn't make sense—none of it. For a corporate holding listed on the New York Stock Exchange, and whose profits had spiked 35 percent since Linsanity's genesis, how believable was it that the true wrench in bringing back the singular spark for that financial windfall would be the question of dollars? If anything, bringing Lin back was supposed to assure money—in advertising, marketing, publishing, untold media avenues—for the

94 Like Bill Walton before him, Yao was unable to overcome persistent plantar fasciitis. As it happens, Persistent Plantar Fasciitis is the name of O'Grady and Cavan's new post-punk acid Celtic band.

Garden coffers from now until Rapture. Even the structure of the proposed deal left more than a few escape hatches.[95] If he struggles for the entirety of the first two years, chalk up years one and two as bad gambles and look to trade what would be arguably one of the most coveted and valuable expiring contracts in league history; worst comes to worst, use the seldom cited "stretch provision" to cut the kid and spread the loss out over the next three seasons. And that's assuming he fails to do what so many young point guards have needed years and years to do: become a serviceable, starting-caliber point guard in the NBA.

At least that's the risk management analysis that most NBA front offices, and most CEOs worth their business school salt, would've run through. But Jimmy Dolan has rarely, if ever, played convention's chord. Self trained in the blues and little else, Guitar Jimmy[96] has for years struggled to show the world that his father, Cablevision founder Charles Dolan, wasn't alone in his service to the bottom line. Forget that the Dolan family is one of the 500 wealthiest on the planet.[97] Forget that the CEO of Madison Square Garden had once signed off on a $30-million-plus contract to a player—Jerome James—who'd never averaged 10 of anything in a game. After treating the luxury tax threshold like a ruby-lipped plastic sex doll for years, this, apparently, was a financial bridge too far.

All of which reeked of Grade-A Barnum & Bailey elephant shit to anyone with a Garden-honed sense of smell. In Dolan's mind, Lin's revised offer sheet amounted to a categorical betrayal—*I, James Dolan, gave you this one last shot, and this is how you repay me—going behind my back?*

If that seems petty and shortsighted, that's because, well, James Dolan is a petty, shortsighted person. No one, save for the most dynastically nepotistic of CEOs, runs a business that way. Nor ought they. Yet somehow, when it came to the Asian American supernova that was Jeremy Lin, even a 35 percent rise in profits wasn't balm enough to cover the wound imagined by the Garden's sole gatekeeper. It didn't matter that Knick management essentially commanded Lin to go out and fetch whatever he could—"set the market"—thus allowing the Knicks to match.[98] Making Lin a preemptive offer made no sense; regardless of any offers made thereafter, New York matching would essentially have amounted to bidding against themselves.

---

95 $5 million or so the first two years, followed by the infamous $15 million year three "poison pill."

96 Yes, the Knicks owner plays guitar and sings in his own band, "JD and the Straight Shot." Hey, you know if you had all that money, you'd pay people to pretend you're a rock star too.

97 The family's cumulative fortune is well in excess of $2 billion. Basically, Dolan boasts a license to print money with his own fat face engraved right on it.

98 Under the rules of the collective bargaining agreement (CBA), teams are allowed to go over the salary cap to re-sign their own players.

That part of the strategy made sense. What made less than no sense was the arrogance with which the Knicks met Houston's initial offer of four years, $28 million,[99] telling anyone that would listen–and before Lin had so much landed in Houston to commence the actual negotiations–that they'd match any offer, and especially *this* one. To which Houston responded precisely the way you'd expect: by upping the offer. The new deal, which will doubtless go down in NBA annals as one of Daryl Morey's finest, most focused managerial maneuvers, included a much larger, more potentially poisonous back-loaded pill in the final year of a condensed three-year contract length.

That all went down on July 14. Despite Houston's chicanery, it was still widely believed that the Knicks would, in Marc Stein's estimation, "match any offer up to a billion dollars." But really, the writing on the wall had been thrown into a higher, near-final relief three days before. Even before rendering their final judgment on the Lin Question, the Knicks moved to bring former point guard Raymond Felton back into the fold. The price: Jared Jeffries, Dan Gadzuric's non-guaranteed contract, recent second round draft pick Kostas Papanikolaou, and a few million Dolan Dollars. Already committed to a 39-year-old Jason Kidd to the tune of three years and nearly $10 million, doubling down on Felton seemed as sure a sign as any that the Knick front office was seriously considering a life after Lin.

Despite Felton's middling career and somewhat polarizing persona,[100] the move did make some semblance of sense on its face; the train long ago hitched to the textbook wrist of Carmelo Anthony, it was understandable that Mike Woodson would look to surround his mercurial cornerstone with guys like Felton, a veteran guard who admittedly played quite well in Mike D'Antoni's speedball offense during the early part of the 2011 season. Where Lin excelled–and thrilled–with the ball in his hands, Felton could reasonably be expected to take the foot off the pedal long enough to let Anthony operate where he felt most comfortable, while remaining a serviceable threat from the perimeter.

But while Felton had indeed performed admirably during his last stin in the orange and blue, his subsequent drop-off–exacerbated by equal parts pouting and awful eating habits–left fans of the once again re-calibrating Knicks understandably nervous. Were we really about to jettison a once-in-a-lifetime phenomenon for an ill-fitting retread, a star in the making for one whose potential had seemingly been long spent? At 28, had Felton already reached his ceiling? Even at $2 million less for the next two seasons, was Felton *that* much of a bargain?

---

99 This offer included a much more palatable "poison pill" of just over $9 million in the final year.

100 He famously feuded with Portland coach Nate McMillan after being traded for Andre Miller shortly before the start of the season. Which makes him roughly the 794th player to feud with Nate McMillan.

To understand the rationale behind the Felton deal, it's important to get a sense of how the curious mind of James Dolan "works."[101] As has been highlighted in a number of lengthy profiles of the Knicks' legacy chief,[102] Dolan—a true and volatile Irishman through and through—above almost all else values loyalty amongst his charges. It's why, despite evidence to the contrary borne on and off the court, he continues to count Isiah Thomas as a confidant, and why so many others have found themselves on the short end of their stubble-bearded boss's vindictive stick. Ray Felton may have never taken a bullet or runaway car to save his check-writer's life, but you can be sure that Dolan remembered how sincerely the North Carolina product lamented his leaving the Big Apple. Through awkward alma mater tussle with fellow Tar Heel Ty Lawson and the aforementioned McMillan beef, Felton clearly yearned to back beneath the bright lights and panoptical eye of the World's Most Famous. When the possibility of letting Lin go became a real one, you can bet that Felton wasn't far from Dolan's thoughts. Which is as terrifying as it is mystifying.

* * *

The day the Knicks made Jeremy Lin's cutting adrift official, James Dolan made an unprecedented appearance at the Las Vegas Summer League during one of the Knicks' afternoon games. Flanked by General Manager Glen Grunwald, Coach Mike Woodson, and Amar'e Stoudemire, a noticeably disheveled Dolan looked on as his depleted squad—cut down nearly in half thanks to the Marcus Camby sign-and-trade 10 days previous—continued its march towards a winless week in Sin City. To many, the showing reeked of opportunistic face-saving, a way to show fans that the looming Lin judgment amounted to a path that had to be taken for the good of the franchise. By merely showing up, Dolan played the bleacher-seated middle finger to Knicks fans everywhere, and served notice that any hope of a changing of the Garden guard was one best left to elaborate acid trip or idle fantasy.

Around the time of Dolan's appearance, word spread that while the Knicks were all but assured in their decision to cut Lin adrift, they wouldn't release an official statement

101 These quotation marks weigh 75 gigabytes each.

102 S.L. Price's 2007 *Sports Illustrated* feature being the subject's Rosetta Stone.

until much closer to the midnight deadline. When official word did finally break around 11 p.m., the reaction on Twitter in particular was one of resigned rage—just because you know something's coming doesn't make you any better prepared to absorb the blow. You could say I didn't take the news well. Another way of saying it would be that I drank heavily and yelled at the neighbors for parking on the edge of my lawn. Yet another way of saying it would be that I was nearly kicked out of my own house for reacting to a basketball player leaving my team with all the grace and dignity of a Jersey shore cuckold finding out the truth in a Jack Daniels haze.[103] Still another way of saying it would be that I tried to summon tears thinking of everyone I've known that's ever died, so I could say from thereon forward that not re-signing Jeremy Lin "made me cry."

Whatever the weird coping mechanism, I certainly wasn't alone. Everywhere you turned, stories of grown men kneeling en sorrowful mass in city bars, of whole families giving away their #17 jerseys to pizza delivery boys, of scorned girls dumping their boyfriends over ill-timed, too-soon jokes, filled the collective zeitgeist...OK, none of that happened. But believe me when I say that peeps were really, really, genuinely upset. Of course they were! After more than a decade of misguided contracts, rudderless front office leadership, truck sex, and shitty basketball *this* is where James Dolan chooses to draw the fiscal line in the sand? After raising ticket prices nearly 50 percent, and top-to-bottom renovations whose pocketbook effects on customers still have yet to be fully felt, this is your Waterloo, Jimmy?

After months of seeming quiet and a front office demeanor that suggested a shift (albeit slight) in organizational culture, here was James Dolan, for years now resigned to spending his public capital behind his Les Paul, surfacing from his self-imposed PR bunker long enough to cast a shadow which even in the cauldron of a record-breaking summer spelled, for Knick Knation at least, an imminent winter. "Lest you forget, this is my team," he seemed to cackle. "Now check out this kick-ass song!"

It will likely be years before we'll be able to quantify the full franchise-altering effects of letting Jeremy Lin walk west into a Texas sunset. For all the magic and beauty and batshit craziness of Linsanity, his is still a career that could veer in any number of wildly disparate directions. Such is the nature of a sport where success is so intertwined with chance and human circumstance that a clearly talented basketball player like Jeremy Lin can fly for years under the radar. Could he be a key piece on a championship team? Perhaps. Might he regress to the role of solid backup? Maybe. Can he be the next Steve Nash? It's possible. Could he bust completely? Sadly, pathetically, and for our own sanity, that's what we're forced to bank on. It's the kind of twisted logic you learn to talk yourself into with the New York Knicks, a team unrivaled in its ability to turn fine wine to piss and one-in-a-million fortune into nothing.

103 There's a story here. We're just not sure what it is.